Environmental Sociology

This new edition of John Hannigan's widely-known and respected text has been thoroughly revised to reflect major recent conceptual and empirical advances in environmental sociology. Key updates and additions include:

- an extended discussion of how classic sociological theory relates to contemporary environmental sociology;
- a focus on cultural sociologies of the environment, notably discourse analysis and social framing;
- updated coverage of the environmental justice movement and global biodiversity loss;
- a critical overview of contemporary interdisciplinary perspectives, namely co-constructionist theories of 'socionature'.

The new edition includes two new 'hot topic' chapters:

- 'Discourse, power relations and political ecology' deals specifically with discursive conflicts between North and South, and includes a profile of contemporary struggles over water privatisation in Africa and Latin America;
- 'Towards an "emergence" model of environment and society' introduces a new way of conceptualising the environmental field that brings together insights from complexity theory, the sociology of disasters, collective behaviour and social movements, perspectives on 'social learning and the sociology of environmental 'flows'.

Written in a lively and accessible manner, *Environmental Sociology* makes a strong case for placing the study of emergent uncertainties, structures and flows central to a 'realist/constructionist model' of environmental knowledge, politics and policy-making. The book offers a distinctive and even-handed treatment of environmental issues and debates, integrating European theoretical contributions such as risk society and ecological modernisation with North American empirical insights and findings.

The book will interest environmental professionals and activists, and will be an invaluable resource to undergraduate and postgraduate students in geography, sociology, political science and environmental studies.

John Hannigan is Professor of Sociology at the University of Toronto and author of two major books, *Environmental Sociology: A Social Constructionist Perspective* (1995) and *Fantasy City: Pleasure and Profit in the Postmodern City* (1998), both published by Routledge.

Environmental Sociology

Second edition

John Hannigan

LONDON AND NEW YORK

First published 1995
by Routledge
2 Park Square, Milton Park, Abingdon, Oxon OX14 4RN

Simultaneously published in the USA and Canada
by Routledge
270 Madison Avenue, New York, NY 10016

Second edition 2006

Routledge is an imprint of the Taylor & Francis Group, an informa business

Typeset in Perpetua by Bookcraft Ltd, Stroud, Gloucestershire
Printed and bound in Great Britain by TJ International Ltd, Padstow, Cornwall

British Library Cataloguing in Publication Data
A catalogue record for this book is available from the British Library

Library of Congress Cataloging in Publication Data
A catalog record for this book has been requested

ISBN10: 0-415-35512-5 (Hbk)
ISBN10: 0-415-35513-3 (Pbk)
ISBN13: 9-78-0-415-35512-4 (Hbk)
ISBN13: 9-78-0-415-35513-1 (Pbk)

To Ruth

Contents

Illustrations

Figures

Tables

Preface

It has been ten years since the inaugural edition of *Environmental Sociology* was published by Routledge. In the ensuing decade, the field of environmental sociology has flourished. Today, the subject is taught at universities across Europe, North America and Australia, as well as establishing a firm foothold in Brazil, Japan and Korea. With growth has come increasing theoretical maturity.

When I started writing the original draft of this book in the early 1990s, environmental scholars were still by and large committed to a strong 'realist' position whereby the gravity, configuration and causes of ecological damage and destruction were thought to be obvious. The recommended role of the environmental sociologist was to expose underlying mechanisms, for example the 'treadmill of production', and map out optimal directions for change. By contrast, a handful of researchers, mostly in the sociology of science, recognised that environmental risks and knowledge were by no means self-evident, but rather the product of social definition and construction. I was inspired to write an entire volume about this after listening to a keynote address by the late Fred Buttel about the socially constructed character of global environmental change (as it happens, Fred later moved away from social constructionism, partly because he feared that it would interfere with the progress of environmental reform).

In recent years, the tone of the realist–constructionist conflict has moderated and environmental sociologists have put their energies into seeking some type of synthesis. In recognition of this, I have deliberately dropped the original subtitle of the book ('A Social Constructionist Perspective'). While I have retained the original emphasis on the sociology of social problems (especially in Chapters 5–9), I have also added quite a bit of new material.

In Chapter 1, you will find a new section in which I describe and assess the contributions to environmental thinking of the trinity of classical sociological theorists: Durkheim, Weber and Marx. In Chapter 2, I offer an extended appraisal of two leading theories of 'the environmental state': Beck's 'risk society' thesis and Mol and Spaargaren's version of 'ecological modernisation theory' (these were surveyed only briefly in the final chapter of the first edition). Furthermore, I have added an overview of the most significant, if not wholly satisfying, attempt to bridge the 'nature–society divide', what is variously known as 'co-constructionism', or in its more specific version, 'Actor–Network Theory' (ANT). Chapters 3 and 4 deal with an aspect of constructionism which was underplayed in the first edition – environmental discourse. Since then, a host of environmental social scientists, most notably Robert Brulle, Eric Darier, John Dryzek and Maarten Hajer, have elevated this to a position of importance. In Chapter 4, I deal specifically with the link between environmental discourse and power, as reflected particularly in contemporary North–South relations. The latter half of this chapter profiles one of the most important environmental conflicts today – water privatisation. Finally, in the concluding section (Chapter 10), I introduce a new perspective, which I label as the 'emergence approach to environmental sociology'. This brings together threads from various fabrics in the environmental literature – social learning, risk society, hybrid management and boundary organisations, globalisation and environmental flows, and collective behaviour in environmental emergencies and disasters – around a focus on uncertainty and indeterminacy, improvisation, adaptation and fluidity.

Acknowledgements

As is the custom, I would like to thank a number of people who have helped make this second edition of *Environmental Sociology* possible.

Among my academic colleagues, Harris Ali (whose path-breaking research on SARS and the Global City I discuss in the final chapter) has been especially helpful. At our occasional lunches over the past few years, he has provided some especially shrewd commentary on the state of environmental sociology and environmental health. Ray Murphy has been a beacon of creative synthesis, lighting up an important way forward with his analysis of the 1998 ice storm and how this may be conceptualised by employing a 'constructionist/realist' approach. Finally, my 2005 graduate seminar on Environmental Sociology provided an excellent opportunity to revisit past material and delve into more recent work.

Many thanks are due to Gerhard Boomgaarden, Senior Sociology Editor at Routledge. While I had been intending to undertake a revision of this book for some time, a combination of heavy university administrative responsibilities and a greater-than-expected success of my 1998 'urban' book *(Fantasy City)* meant that I kept putting this off indefinitely. Gerhard's enthusiasm for the project convinced me that now was the time. Also at Routledge, Constance Sutherland has been a helpful and patient liaison, ensuring that I would more or less meet the required deadline.

As was the first edition, this book is dedicated to my wife Ruth. Without her inspiration and love, 'ES 2' (as it became known around here) would never have got off the ground. I especially value our conversations about the state of the world on Sunday mornings over coffee/tea and the *New York Times*. Tim, our younger son, spent many hours attending to computer-related matters, including converting me into a 'wireless' user. Home from university for the summer, Maeve, our eldest daughter, typed a significant portion of the super-sized bibliography. The other two Hannigan children, T.J. (who helped in the computer set up and printing of the first edition) and Olivia (who will almost certainly be dragooned into service on the next one) were also supportive. Thanks to everyone.

Abbreviations

AEC	Atomic Energy Commission (USA)
Alpac	Alberta–Pacific Forest Industries
ANT	Actor–Network Theory
AOU	American Ornithologists Union
AVIVE	Green Life Association of Amazonia (Brazil)
CBC	Canadian Broadcasting Corporation
CBD	Convention on Biological Diversity
CCB	Center for Conservation Biology (Stanford University)
CFCs	Chlorofluorocarbons
CIPRs	Collective Intellectual Property Rights
CITES	Convention on International Trade in Endangered Species of Wild Fauna and Flora
CMS	Convention on Conservation of Migratory Species of Wild Animals
DoE	Department of the Environment (UK)
DRC	Disaster Research Center
DSP	Dominant Social Paradigm
EAAB	Empresa de Acueducto y Alcantarillado de Bogotá (Colombia)
EJM	Environmental Justice Movement
EJO	Environmental Justice Organisation
ENGO	Environmental Non-Governmental Organisation
EPA	Environmental Protection Agency (USA)
EWG	Environmental Working Group (USA)
FDA	Food and Drug Administration (US)
GREEN	Genetic Resource, Energy, Ecology and Nutrition Foundation (India)
HEP	Human Exemptionalism Paradigm
INBIQ	National Biodiversity Institute (Costa Rica)
IPCC	International Panel on Climate Change
IRI	International Research Institute for Climate Protection
IUCN	International Union for Conservation of Nature
JAES	Japanese Association for Environmental Sociology
NAACP	National Association for the Advancement of Colored People (USA)
NAS	National Academy of Sciences (USA)
NASA	National Aeronautics and Space Administration (USA)

NEP	New Ecological Paradigm
NGO	Non-Governmental Organisation
NOAA	National Oceanic and Atmospheric Administration (USA)
NSF	National Science Foundation (USA)
NSMs	New Social Movements
ONAC	Office of Noise Abatement and Control
OPEC	Organisation of Petroleum Exporting Countries
PVPA	Plant Variety Protection Act (USA)
SARS	Severe Acute Respiratory Syndrome
SBSTA	Subsidiary Body for Scientific and Technological Advice
SCB	Society for Conservation Biology
SMO	Social Movement Organisation
UCC	United Church of Christ
UFW	United Farm Workers
UNCED	United Nations Conference on Environment and Development
UNEP	United Nations Environment Programme
UNESCO	United Nations Economic, Social and Cultural Organisation
UNFCCC	United Nations Framework Convention on Climate Change
USAID	US Agency for International Development
USDA	US Department of Agriculture
WWF	Worldwide Fund for Nature

1 Environmental sociology as a field of inquiry

'Earth Day 1970' is often said to represent the debut of the modern environmental movement. Starting as a modest proposal for a national teach-in on the environment, it grew into a multi-faceted event with millions of participants. What most distinguished Earth Day, however, was its symbolic claim to be 'Day 1' of the new environmentalism, an interpretation which was widely embraced by the American mass media, thus affording the environmental issue instant and widespread recognition (Gottlieb 1993: 199).

When Earth Day inaugurated the 'Environmental Decade' of the 1970s, sociologists found themselves without any prior body of theory or research to guide them towards a distinctive understanding of the relationship between society and the environment. While each of the three major classical sociological pioneers – Émile Durkheim, Karl Marx and Max Weber – arguably had an implicit environmental dimension to their work, this had never been brought to the fore, largely because their American translators and interpreters favoured social structural explanations over physical or environmental ones (Buttel 1986: 338). From time to time, isolated works pertaining to natural resources and the environment had appeared, mostly within the area of rural sociology, but these had never coalesced into a cumulative body of work. In a similar fashion, social movement theorists gave short shrift to conservation groups, leaving historians to explore their roots and significance.

To comprehend why this situation arose, it is necessary to consider how both geographical and biological theories of social development and social change lost their predominance when sociology emerged as a distinctive discipline in the early twentieth century.

The failure of geographical and biological determinism

In the nineteenth century, the effects of the geographical environment on the human condition was a topic of considerable scholarly interest. Perhaps the leading geographical determinist was the British historian, Henry Thomas Buckle, author of *The History of Civilization in England*. Buckle was greatly influenced by the writing of the seventeenth-century French philosopher, Montesquieu, and by several German geographers, notably Karl Ritter. His central thesis was that human society is a product of natural forces, and is therefore susceptible to a natural explanation (Bierstedt 1981: 2). Buckle believed that the influence of the geographical environment is most direct and therefore strongest upon 'primitive' people but declines with the advance of modern culture. He ascribed particular sociological significance to the visual aspect of nature: if the natural environment is awe-inspiring in its beauty or terrifying in its power of destruction, it overdevelops the imagination; if it is less formidable, a more rational intelligence prevails. England, with its gently rolling hills and domesticated farm animals, represented a prime example of the latter.

Buckle's geographical theory of social change was widely read and quite influential in intellectual circles in the nineteenth century (Timasheff and Theodorson 1976: 93). For example, the economist Thomas Nixon Carver used *The History of Civilization in England* in his sociology course at Harvard long before that university had a formal department of sociology, while William Graham Sumner, widely regarded as the first American sociologist, became interested in Buckle's work while studying theology at Oxford (Bierstedt 1981: 2).

A second leading geographical determinist was Ellsworth Huntington. In his principal sociological works, *Civilization and Climate*, *World Power and Evolution* and *The Character of Races*, Huntington attempted to establish a series of correlations between climate and health, energy, and mental processes such as intelligence, genius and willpower. Having divined the parameters of an 'optimal climate' he then attempted to prove that the rise and fall of entire civilisations such as that of ancient Rome follow the shift of the climatic zones in historical periods.[1]

In assessing the worth of this 'geographical school', Sorokin (1964 [1928]: 192–3) refers to its fallacious theories, its fictitious correlations and its overestimation of the role of the geographical environment, but at the same time he cautions that 'any analysis of social phenomena which does not take into consideration geographical factors is incomplete'.

The natural world also entered into early sociological discourse through the Darwinian concepts of 'evolution', 'natural selection' and the 'survival of the fittest'. In Darwin's theory, those plants and animals which are best suited to adapt to their environment survive, while those which are less well equipped perish. The survivors pass on their advantages genetically to subsequent generations. Darwinism was seized upon by many of the early conservative sociological thinkers who applied its principles (not always accurately) to the human context (see Hofstadter 1959). The most prominent social Darwinist was the English social philosopher, Herbert Spencer, who proposed an evolutionary doctrine which extended the principle of natural selection to the human realm. Spencer bitterly opposed any suggestion that society could be transformed through educational or social reform; rather, he believed that, if left alone, progress would evolve in a gradual fashion.

Sumner was Spencer's greatest academic disciple in America, introducing his own concept of the 'competition of life' whereby humans struggle not just with other species for survival in the natural universe but also with each other in a social universe. Applying his theory to the *laisser-faire* capitalism of the day, Sumner legitimated the triumph of the 'robber barons', millionaire industrialists who made their money in banking, railroads and utilities through sharp and ruthless dealing. They were, Sumner claimed, 'a product of natural selection' who would move society forward on the road to progress.

Both these 'single factor theories of social change' (Bierstedt 1981: 487) were rejected by mainstream sociology for largely the same reasons. By the 1920s the evolutionary *laisser-faire* doctrines of the nineteenth century had given way to a new emphasis on social planning and social reform. 'Meliorism' – the deliberate attempt to improve the well being of members of society – flew in the face of these social theories which viewed social causation as unalterable, whether due to geography or biology.

Furthermore, by this time the foundation of sociological theory had shifted. Many sociologists had come to accept psychology as the foundation of sociology in place of physics or biology (Timasheff and Theodorson 1976: 188). This was especially evident in the social psychological tradition established by Mead, Cooley, Thomas and other American 'symbolic interactionists' who emphasised that the reality of a situation lies entirely in the definition attached to it by participating social actors. This definition, in turn, was socially shaped, as in Cooley's concept of the 'looking glass self'. Physical (and environmental) properties became relevant only if they were perceived and defined as relevant by the actors (Dunlap and Catton 1992/3: 267).

Increasingly, the failure of social Darwinism, and to a lesser extent the inability of geographic determinism to ever get off the ground, led to a strong aversion to explanations which used biological–environmental explanations. This opposition to biological currents was similarly evident in sociology's sibling discipline, anthropology.

After his move to the United States in 1886, Franz Boas, widely recognised as the founder of American cultural anthropology, responded to the rising tide of eugenics, 'scientific' racism and other manifestations of biological determinism by elevating culture to a primary role in individual and societal development, dwarfing both the physical environment and biological inheritance. This emphasis on cultural processes was carried on in the twentieth century by such well-known anthropologists as Margaret Mead and Ruth Benedict (Benton 1991: 13). Culture, in fact, came to be valued as the key influence on all aspects of human society.

Ironically, while sociology rid itself of biological explanation, it hung on to a distinctly biological terminology. Functionalism, the leading sociological theory of the 1950s in America, carried forward Durkheim's idea that society constituted a social 'organism' which was constantly having to adapt to the outside social and physical environment. Its equilibrium or steady state could be knocked out of kilter by various disruptive events but, ultimately, it would return to normal just as the human body recovers from a fever. Dickens (1992) has noted that functionalist theorists, especially their dean, Talcott Parsons, might have gone further and actually developed a theory of social evolution in an environmental context which stressed how biological inheritance permitted humans both to adapt to the natural world and to change it. This potential, however, was never developed, leaving environmental factors as marginal elements in sociological explanation.

Sociologists as 'hucksters' for development and progress

A second explanation for sociological foot-dragging on environmental matters pertains to the world-view of sociologists themselves.

In a steady stream of papers and articles from the late 1970s on, William Catton and Riley Dunlap argued that the vast majority of sociologists share a fundamental image of human societies as exempt from the ecological principles and constraints that govern other species. While sociologists are inclined to favour the use of social engineering to achieve such goals as equality, they nevertheless fully accept the possibility of endless growth and progress via continued scientific and technological development while ignoring the potential constraints of environmental phenomena such as climate change (Dunlap and Catton 1992/3: 270).

Some sociological specialities went even further, actively becoming advocates, and even 'hucksters', for the benefit of technological innovation and economic development. Nowhere was this more evident than in the sociological literature on modernisation which was influential for two decades between 1955 and 1975.

Two works in particular stand out in the study of the modernising process: Inkeles and Smith's *Becoming Modern* (1974) and Lerner's *The Passing of Traditional Society* (1958).

For Inkeles and Smith, modernisation denotes both a societal and personal transformation. At the societal level, modernisation is conceptualised as a process of nation and institution building. In the 1960s, the 'decade of development', many Third World nations failed to make their entry into the modern world, sliding backwards into tribalism and ethnic conflict. Newly liberated from colonialism, these emerging countries were said to be 'hollow shells, lacking the institutional structures which make a nation a viable and effective socio-political and economic enterprise' (Inkeles and Smith 1974: 3).

Inkeles and Smith argue that the primary reason for this failure to modernise was that individual members of the community were psychologically trapped in the past, unable to transcend traditional ways of thinking to become modern personalities. Modern citizens possessed a panoply of skills: they could keep to fixed schedules, observe abstract rules, adopt multiple roles and empathise with others. They were optimistic, opinionated, open to new experience and consumers of information. These qualities are not inborn but must be acquired through life experience.

While some of this socialisation in modern ways could be carried out by the educational system, it is the factory, Inkeles and Smith conclude, that is the true 'school in modernity'. The factory, they observed, is the epitome of the institutional pattern of modern civilisation. It functions as a powerful model for rural migrants from traditional settings inculcating, among other qualities, a sense of efficacy, a readiness for innovation and an openness to systematic change, respect for subordinates and the importance of planning and time.

To Daniel Lerner (1958), the key correlate of developing modernity is the media's role in establishing a psychological openness to change among peasant populations. In particular, the media were depicted as fostering a sense of 'empathy' – the ability to imagine change by putting oneself in the shoes of those in society who were engaged in playing roles (e.g. social leader) other than one's own.

In the ascent to modernity the influence of the physical environment was downgraded. Inkeles and Smith (1974: 22) observed that a key part of developing a sense of modern efficacy lay in the ability to develop a potential 'mastery' of nature. In their questionnaire, administered to a thousand men in six developing countries, they pose this question:

> Which of the following statements do you agree with more?
>
> 1 Some people say that man will some day fully understand what causes such things as floods, droughts and epidemics.
>
> 2 Others say that such things can never fully be understood by man.

The respondent who was more committed to advancing his own goals rather than being dominated by natural forces would respond positively to the first statement. This view of nature is of course the antithesis of the ecological ethic which stresses that human beings have no inherent claim to domination over nature but must simply coexist with other species on the earth.

One of the few commentators on modernisation in the 1960s to recognise the potential constraints imposed by the environment was Clifford Wharton, an agricultural economist, who noted the special characteristics of agriculture which related to climate, soil and other inputs. 'Bananas do not grow in Alaska (except perhaps in a hothouse)' but 'a shoe factory in Tokyo need not be different from one in São Paulo', Wharton (1966) observed, concluding that agriculture was far more subject to environmental factors than other forms of economic development.

Mesmerised by the benefits of economic development and its sidekick, individual modernity, most sociologists, by contrast, either completely ignored the natural environment or viewed it as something to be overcome with grit and ingenuity.

That is not to say that there were not isolated critics of the pro-development paradigm, especially within the ranks of Marxist sociology. But, like religion, they tended to see the environment as a distraction from the necessity of class struggle. Even where the seriousness of environmental destruction was acknowledged, left-wing critics were inclined to focus on the class and power relations underlying this crisis rather than on factors relating more directly to the environment itself (see Enzenberger 1979). Insomuch as Marxism eventually came to dominate social theory in some important regions of post-war European social theory, this resulted in the further exclusion of environmental issues from the discipline of sociology (Cotgrove 1991; Martell 1994).

Classical sociological theory and the environment

One possible source of inspiration for contemporary sociologists seeking to engage with environmental topics is the canon of classical social theory, notably that bequeathed to us by Durkheim, Weber and Marx. To a certain extent each of these sociological pioneers had something significant to say about nature and society, although this was often more implied than direct, and was embedded in the philosophical controversies and scholarly debates of the time in which they were writing.

Some commentators have been decidedly downbeat about the potential usefulness of this canon. Goldblatt (1996: 1–6), for example, advises that we be wary of the legacy left to us by classical sociological theory insofar as it lacks an adequate conceptual framework with which to understand the complex interactions between societies and environments. Rewarding though it may be, Järvikowski (1996: 82–3) says, the reading of classic works by this triumvirate is simply not sufficient for adequate theorizing of contemporary environmental problems. Finally, Buttel (2000: 19) concludes that the legacy bequeathed by classical sociology is very much mixed: some of the tools initially developed by the classical theorists are needed, but 'the overall thrust of the classical tradition was to downplay ecological questions and biophysical forces'.

On the other hand, there is a rich and expanding corpus of work in which environmental scholars seek to reveal this conventional wisdom to be premature. As we will see, some commentators (William Catton, John Bellamy Foster) deliberately adopt the strategy of extracting 'ecological' insights from the work of the classic thinkers that have been overlooked or misunderstood in the past. Others (Raymond Murphy, Peter Dickens) are more inclined to smoke out concepts and ideas from the collected works of the sociological pioneers, even if these were not originally used in an environmental context, and apply them to the current environmental 'crisis' with some intriguing results. Some analysts have chosen to adopt a typological approach, organising the field on the basis of classical theory. For example, Sunderlin (2003) defines and conceptualises three key paradigms (individualist, managerial, class), each of which is derived from the classical sociological literature (Durkheim, Weber, Marx).

Émile Durkheim

Of the three founding figures in sociology, Durkheim is probably the least likely to be recognised as an environmental commentator.[2] In large part, this reflects his deliberate decision to elevate *social facts* over 'facts of a lower order' (that is, psychological, biological).

For Durkheim, a social fact is 'any way of acting, whether fixed or not, capable of exerting over the individual an external constraint' (2002 [1895]: 117). This constraint is normally manifested in the form of law, morality, beliefs, customs and even fashions. We can verify the existence of a social fact, Durkheim ventured, by examining an experience that is characteristic. For example, children are compelled to adopt ways of seeing, thinking and acting that they otherwise would not have arrived at spontaneously.

Durkheim is quite firm in asserting that social phenomena cannot be explained through the lens of individual psychology. It is a central rule of the sociological method that 'the determining cause of a social fact must be sought among antecedent social facts and not among the states of individual consciousness' (p. 125). This rule may infuriate strong advocates of individualism, but no matter. Social facts, Durkheim insists 'are consequently the proper field of sociology' (p. 112).

While this vigorous defence of social facts and collective consciousness most certainly buttressed the theoretical independence of sociology, it also had the effect of warning off members of the new discipline from non-sociological approaches that were reductionist in nature (that is, they reduced explanation to biological or psychological factors).

Nevertheless, Durkheim himself frequently utilised biological concepts and metaphors in presenting his theory of societal transformation. Furthermore, this theory was most certainly inspired by the Darwinian evolutionary model that was popular among intellectuals in the late nineteenth century. In *The Division of Labour in Society* (1893), he describes the evolution of modern societies from a state of *mechanical solidarity*, wherein social solidarity is a product of shared cultural values, to one of *organic solidarity*, where the social bond is a function of interdependence, most notably that arising out of an increasingly complex division of labour.

Catton (2002: 92) proposes that Durkheim's theory was very much an attempt to devise a solution to what is essentially an ecological crisis of rising population paired with scarce resources. As societies became larger and denser, it would have been disastrous if everyone had continued to engage in agriculture. Increasingly, occupational specialisation meant that the competition over arable land was lessened, even as that land became more productive thanks to technological innovation.

Alas, Durkheim was doubly hobbled, Catton says, both by his narrowly selective reading of Darwin and by the unavailability in the 1880s of our knowledge today of ecology and evolution (2002: 93). In the first instance, he erroneously supposed that Darwin believed increasing diversity to be a way of minimising competition for scarce resources. Rather, Darwin cautioned that co-evolution (two species evolving at the same time) could, in some cases, increase their *resemblance* to one another or result in one species bringing the other to extinction. In short, Darwin viewed specialisation as a way in which one species could gain competitive advantage over another, not, as Durkheim believed, as a way of *lessening* rivalries and increasing mutual interdependence. Furthermore, Durkheim could not have been privy to the insights of modern ecology, which did not emerge as a sub-field of biology until the next century. Most crucially, no one in Durkheim's time recognised that mutual dependence was symbiotic but not necessarily balanced. That is, some interactions in nature benefit both member populations (*mutualism*) but others benefit one without either harming or benefiting the other (*commensalism*); and yet others are beneficial to one and detrimental to the other, as with predators and parasites (Catton 2002: 93). The latter gives rise to power differences, something especially significant when you are dealing with human ecological communities.

What we are left with then is chiefly speculation on what *might have been*. Citing Talcott Parsons (1978: 217), Järvikowski (1996: 82) ventures that Durkheim would likely have written in a different way today about the relations between the social and physical environments because biological theory has undergone a profound process of change.

Max Weber

A second sociological pioneer whose work is said to possess an ecologically relevant component is Max Weber. As Buttel (2002) has pointed out, this environmental connection has been located in two entirely different corners of Weber's work by Patrick West and Raymond Murphy.

West (1984) draws mostly on Weber's historical sociology of religion and his comparative research on ancient societies. He emphasises that Weber analysed concrete examples of struggles over natural resources, for example, the control of irrigation systems.

By contrast, Murphy's more extensively drawn discussion of neo-Weberian environmental sociology is based primarily on Weber's book *Economy and Society* (1978 [1922]). For Murphy, the key concept to be extracted here is *formal rationalisation*. Rationalisation is composed of several dynamic institutional components. Increased scientific and technical knowledge brings with it a fresh orientation in which nature exists only to be mastered and manipulated by humans. An expanding capitalist market economy leaves little room for anything beyond the calculating, self-interested pursuit of market domination. Industry and government are controlled by a bureaucratic apparatus, the purpose of which is is to attain a high level of efficiency. The legal system operates like a technically rational machine. Together, these components promote a pervasive logic whereby efficiency reigns supreme, on occasion even superseding a sensible choice of goals or alternatives, what Weber called *substantive rationality*. Formal rationality thus dictates that the most efficient action is to clear-cut an old growth forest, even if this is in no way substantively rational from an ecological point of view (Murphy 1994: 29–30).

Murphy (1994: 34) identifies two interrelated processes first highlighted by Weber at the beginning of the twentieth century that have become distinctive features of our time: the *intensification of rationality* and the *magnification of rationality*. The more we try to run things according to the principle of dispassionate calculation the more we open the door to a swarm of unwanted and negative effects. When applied to the case of nature, this is called *ecological irrationality*. It is manifested in a wide range of destructive consequences from sensational technological disasters such as nuclear accidents to routine pollution events such as industrial dumping into urban storm sewers.

Drawing on another of Weber's (1946 [1918]) concepts – *intellectual rationality* – Freudenburg (2001) makes an important point about science, technology and risk. In contrast to tribal societies, the average individual in an industrial society cannot know more than a minimum about how technology works – unless he or she is a physicist, one who rides on the streetcar has no idea how the car happened to get into motion (Weber 1946 [1918]: 138–9). Consequently, while one may in principle master all things by intellectual calculation, in reality we depend on an army of experts to do so. Yet, as Freudenburg notes, this expectation is inherently problematic because a minority of the time these experts fumble the ball, leading to potential, and sometimes actual, environmental emergencies.

Karl Marx

Of the three main sociological traditions, it is that associated with Karl Marx that has provoked the most extensive response from present-day environmental interpreters. Marx and his early collaborator Friedrich Engels were only marginally concerned with environmental degradation *per se* but their analysis of social structure and social change has become the starting point for several formidable contemporary theories of the environment.

Marx and Engels believed that social conflict between the two principal classes in society, that is capitalists and the proletariat (workers), not only alienates ordinary people from their jobs but also leads to their estrangement from nature itself. Nowhere is this more evident than in 'capitalist agriculture' which puts a quick profit from the land ahead of the welfare of both humans and the soil. As the industrial revolution proceeded through the eighteenth and nineteenth centuries, rural workers were removed from the land and driven into crowded, polluted cities while the soil itself was drained of its vitality (Parsons 1977: 19). In short, a single factor, capitalism, was held responsible for a wide range of social ills from overpopulation and resource depletion to the alienation of people from the natural world with which they were once united. Marx and Engels saw the solution as the overthrow of the dominant system of production, capitalism, and the establishment in its place of a 'rational, humane, environmentally unalienated social order' (Lee 1980: 11).

Marx and Engels argue for the establishment of a new relationship between people and nature. However, it is not entirely clear what form such a relationship should take. In the work of the more mature Marx, this seems to follow a distinctly anthropocentric direction depicting humans as achieving mastery over nature, in no small part because of technological innovation and automation. This has been called a Promethean (pro-technological, anti-ecological) attitude toward nature (Foster 1999: 372; Giddens 1981: 60).

By contrast, in Marx's early work the concept of the 'humanisation of nature' is proposed. This suggests that humans will develop a new understanding of and empathy with nature. A key question here is whether this new understanding would be used solely for human emancipation or whether it would take a more 'ecocentric' form in which the powers and capacities of non-human species would be enhanced. In the former case, the humanisation of nature might, in fact, be deployed to eliminate species and organisms that threaten human health (Dickens 1992: 86). As Martell (1994: 152) observes, the texts of the early Marx are too complicated and contradictory on ecological concerns to be the basis for a fully fledged theory of environmental protection; it may be more useful to pursue this project through other sources or frameworks.

Contemporary Marxist theory emphasises not only the role of capitalists but also that of the state in fostering ecological destruction. Both elected politicians and bureaucratic administrators are depicted as being centrally committed to propping up the interests of capitalist investors and employers. While the incentive here is partly material (e.g. corporate campaign contributions, future job offers), public servants, politicians and capitalist producers are said to share an 'ethic' which accentuates capitalist accumulation and economic growth as the dual engines which drive progress. This, they argue, applies at all political levels from the global system to the local community.

One widely noted reading of Marx's environmental views is John Bellamy Foster's seminal article on Marx's theory of *metabolic rift*. According to Foster, Marx has been wrongly accused of providing little insight into the 'ecological crisis' of our times. Indeed, due to the Promethean attitude that suffuses his later writing he may even have impeded the understanding of environmental problems. To the contrary, Foster argues:

> Marx provided a powerful analysis of the main ecological crisis of his day – the problem of soil fertility within capitalist agriculture – as well as commenting on the other major ecological crises of his time (the loss of forests, the pollution of the cities, and the Malthusian specter of overpopulation). In doing so, he raised fundamental issues about the antagonism of town and country, the necessity of ecological sustainability, and what he called the 'metabolic' relation between human beings and nature.
>
> (1999: 373)

It is this latter issue that Foster addresses most substantively in his article. Borrowing from the vocabulary of mid-nineteenth-century chemistry, Marx employed the concept of metabolism to describe the complex interaction between society and nature. Metabolism, he observed 'constitutes the fundamental basis on which life is sustained and growth and reproduction become possible' (Foster 1999: 383). By the 1860s, this organic relationship was being seriously undercut by the practices of capitalist agriculture. Most notably, landowners were accused of callously robbing the soil of its key nutrients by declining to recycle them. This, of course, is exactly what is still occurring, especially where monocultures (a single variety of a single crop grown for commercial profit) prevail. Marx describes this as a 'metabolic rift'– the estrangement of human beings from the natural world of the soil. This paralleled the estrangement of workers from their labour and was attributable to the same source – capitalism.

Rather than a huckster for chemical agriculture, Marx (and Engels) appears to have been an early advocate of organic farming methods. For example, he writes at length about the benefits of spreading manure on crop lands, even suggesting that human waste from the city be recycled as fertiliser rather than polluting the rivers and oceans. Strangely enough, his inspiration for this view seems to have been the German agricultural chemist Justus von

Liebig, who achieved renown as the inventor of synthetic fertilisers. By the late 1850s, Liebig had evidently come to the conclusion that soil depletion was becoming a major problem, especially in America where vast tracts of arable land were cultivated for the sole purpose of exporting grain to the big cities. Liebig even went so far as to recommend that the city of London organically recycle its sewage rather than dump it in the river Thames.

For Foster, the importance of Marx's theory of metabolic rift lies not just in his repatriation of Karl Marx as an advocate of organic agriculture but also in his successful application of sociological thinking to the ecological realm. Foster (1999: 400) calls this 'one of the great triumphs of classic sociological analysis' and proof that 'ecological analysis, devoid of sociological insight is incapable of dealing with the contemporary crisis of the earth'. Furthermore, it provides a portal through which contemporary environmental analysts might better understand the metabolic relation between humans and nature.

One recent effort along these lines is York *et al.*'s (2003: 36–8) discussion of how the metabolic rift can lead to increases in GHG (greenhouse gas) emissions. Three ways this occurs are specified here: the increased transportation of natural resources necessitated by urbanisation; the replacement of organic matter by chemical fertilisers; and the diversion of methane-generating organic waste to landfill rather than back into the soil.

Towards the emergence of an environmental sociology: 1970–2005

There are various reasons why a new scholarly field appears on the academic horizon. Sometimes this reflects the expanding possibilities bursting forth from a cutting edge methodology or theoretical breakthrough. For example, Crick and Watson's unravelling of the double helix structure of DNA was the catalyst that sparked the growth of cell biology. At other times, a new specialisation represents the merger of two previously existing scientific specialities. Finally, a new field can arise out of the intellectual and political ferment generated by movements for social reform and change. This probably best describes the case of environmental sociology.

As we have seen, each of the three widely acknowledged 'founders' of the discipline of sociology – Durkheim, Weber and Marx – addressed some aspect of nature and society, but this was not really definitive to their work. If environmental interest was to be found anywhere in North America, it was within the area of rural sociology, where there was a body of empirical research on natural resources. These enquiries took two forms: the study of natural resource dependent communities and research on the burgeoning use of public parkland for recreational purposes (Humphrey *et al.* 2003: 11). Alas, by the late 1960s, many of these contributions had been overlooked or totally forgotten (Freudenburg and Gramling 1989: 44).

There is general agreement that the first explicit use of 'environmental sociology' was by Samuel Klausner in his 1971 book *On Man in His Environment* (page 4). Dunlap (2002b: 11–12) remembers that he first came across the term in Klausner's book several years later 'when the term was just starting to be used'. Throughout the 1960s, Klausner, a sociologist and clinical psychologist, was engaged in a series of studies of human behaviour under stress. In 1967, he received a small grant ($7,000) from a think tank, Resources for the Future, to study 'social–psychological aspects of environmental research'. Three years later, he edited a special issue of the *Annals of the American Academy of Political and Social Science* on 'Society and Its Physical Environment'.

By this time, sociological interest in environmental matters had been re-ignited, primarily by the rising popularity of environmentalism and the environmental movement. A major catalyst for this had been the publication a decade before of *Silent Spring* (1962), Rachel Carson's bestselling expose of ecosystem damage due to agricultural pesticide use. Then in the early 1970s, the widespread attention accorded the apocalyptic predictions contained in *The Limits of Growth* (Meadows *et al*. 1972), combined with the 'energy crisis' in the United States, deepened this environmental concern among academics. In addition, it broadened the scope of sociological interest in environmental matters to include issues related to resource scarcities and energy use. One sociologist who was particularly swayed by this was William Catton. Upon his return from New Zealand to the University of Washington in 1972, Catton expanded his earlier research interest in national parks and wilderness visitors to a more theoretical concern with overpopulation and declining fossil fuels. This coalesced with the publication in 1980 of the influential book *Overshoot: The Ecological Basis of Revolutionary Social Change*.

Reflecting back on that period, Riley Dunlap, regarded as one of the founders of the field, identifies a two-step progression. At first, researchers, impressed with the great deal of attention that environmental issues were receiving, applied traditional sociological perspectives on public opinion, social movements and formal organisations to topics such as the social characteristics of environmental activists, and the tactics and strategies employed by environmental groups. Gradually, however, interest shifted towards the establishment of an environmental sociology that might be distinctive enough to warrant having a field in its own right (Dunlap 2002a: 329). The focus here was to be the underlying relationships between modern industrial societies and the physical environments they inhabit (Dunlap and Catton 1979).

To underscore this, some key contributors took pains to distinguish strictly between a *real* 'environmental sociology' that focused on the study of environment–society interaction; and a 'sociology of environmental issues' that did not (Dunlap and Catton 1979; Catton and Dunlap 1978). However, in the decades which followed, this distinction became blurred and environmental sociology now tends to be used simply to describe 'the kinds of work that is conducted by self-identified environmental sociologists' (Dunlap 2002a: 346).

In Europe, stimulated by the emergence of the 'greens' as a political force, most of the early work on environmental topics dealt with environmentalism and the environmental movement (Dunlap and Catton 1992/3: 273). One exception to this was in the Netherlands where nodes of activity in environmental sociology formed early on around questions pertaining to agriculture and risk assessment. In Britain, interest in the environment tended to be explicitly theoretical, weighing the relationship between society and nature against classical sociological perspectives on social class and industrialism. By the 1980s, however, empirical research on environmental topics began to flourish in the UK, in part due to the stimulus provided by the Global Environmental Change Programme set up by the Economic and Social Research Council (ESRC), which underwrote an impressive array of conferences, study groups and symposia.

Environmental sociology has also been established since the early 1990s in Japan and Korea. One of the first environmental researchers in Japan was Nobuko Iijima who wrote her Master's thesis on the impact of Minimata disease on the local community. In 1992, she helped found the Japanese Association for Environmental Sociology (JAES) and served as its first president (Hasegawa 2002). By 1999, the JAES had over 450 members and its own publication, the *Journal of Environmental Sociology* (*Kankyo Shakaigaku Kenkyu*) [Mitsuda and Fisher 2000]. In Korea, environmental sociology began to be taught from the early 1990s. Following a 1993 international conference held under the title 'Environment and Development', the Research Group for Environmental Sociology was established in 1995. This led to the founding of the Korean Association for Environmental Sociology in June 2000 (Lee and Park 2002). In October 2001, at the Kyoto Environmental Sociology Conference, a research network, the Asian Pacific Environmental Connection was founded with the brief of solving societal and environmental problems in the Asia–Pacific region (Mitsuda 2002).

After three decades of activity, the state of environmental activity today is both encouraging and disappointing.

In their introduction to the *Handbook of Environmental Sociology*, Riley Dunlap and his co-authors (William Michelson and Glen Stalker) note the 'diversity' and 'richness' of sociological work dealing with the physical environment. This observation is most certainly accurate. There are at least nine distinct competing paradigms: human ecology, political economy, social constructionism, critical realism, ecological modernisation, risk society theory, environmental justice, actor–network theory and political ecology. At the same time, the theoretical repertoire of environmental sociology has been reasonably resistant to the danger of drifting into excessive pluralism or theoretical disarray, maintaining a surprising degree of continuity. Indeed, most of the empirical issues of interest to environmental sociologists today are the very same as those that commanded attention in the past:

> the nature of environmental social movements; states, politics and environmental policy formation; environmental attitudes, beliefs and values; the relationships between consumption and production institutions; the reciprocal impacts of societies and environments; the role of technology in social and environmental change; and the significance of 'the global' in terms of 'environmental scale' and social institutions.
>
> (Buttel *et al*. 2002: 28)

One indicator that a speciality area is becoming firmly rooted in the academic firmament is the establishment of separate sections and research committees within professional associations. One of the earliest instances of this was the formation of the Section on Environment and Technology within the American Sociological Association (ASA) in 1976. A quarter-century later, the section had 409 paid-up members, ranking it number 23 of ASA's 43 sections and sections in formation (Lewis and Humphrey 2005: 154).

Outside North America, one of the most active organisational vehicles for environmental theory and research has been the Research Committee on Environment and Society (RC 24) within the International Sociological Association (ISA). This is the product of a 'merger' in 1992 between RC 24 and the former ISA Research Committee on Social Ecology (RC). Presidents of RC 24 have included Riley Dunlap, Fred Buttel and Arthur Mol. It has long been the Committee's policy to sponsor conferences and other fora, both at the quadrennial World Congress of Sociology and in the years in between. Conferences of the latter type have been held in the Netherlands, Brazil, France, Japan and the United States. Several of these have culminated in edited volumes that highlight recent theoretical developments in environmental sociology (Dunlap *et al.* 2002; Spaargaren *et al.* 2000).

Over the past decade, two separate 'handbooks' of environmental sociology have been published. *The International Handbook of Environmental Sociology* (1997) has as its senior editor Michel Redclift, a respected British scholar in the sociology of agriculture and international development studies. It tends to favour a left-leaning perspective. *The Handbook of Environmental Sociology* (2002) is more North American in orientation. Reflecting the differing interests of its editors, Riley Dunlap and William Michelson, it distinguishes between the 'natural' and 'built' environments, with contributors being drawn from both streams of research. As Buttel (2002: 48) has reported, notable examples of entries in discipline-wide compendia and encyclopaedias include his paper (Buttel and Gijswijt 2001) in the *Blackwell Companion to Sociology*, Dunlap and Rosa's (2000) piece in the *Encyclopedia of Sociology* and entries authored by Mertig and Dunlap (2003) and Schnaiberg (2003) in the *International Encyclopedia of the Social and Behavioral Sciences*. Environmental sociology has been well represented in the *Annual Review of Sociology*, with state-of-the-art chapters appearing on three occasions since 1979 (see Dunlap and Catton 1979; Buttel 1987; Goldman and Schurman 2000).

There is some evidence that environmental sociology has been making inroads into publishing and teaching in mainstream sociology, although it is by no means as influential as long-established specialities such as deviance, stratification and demography. In one study from the mid-1990s, Krogman and Darlington (1996) surveyed mainline, refereed sociology journals from 1969 to 1994. Their data indicate that the number of environmental articles appeared to be significantly increasing over the latter years (total = 75 between 1990 and 1994, as compared to 36 between 1985 and 1989). A decade later, Lewis and Humphrey (2005) reported the results of a content analysis of 24 widely adopted American introductory sociology textbooks. On average, the texts cited four works by influential environmental sociologists. This positive news was tempered, however, by their finding that the texts omitted some of the most central, unique concepts in the field and typically treated environmental issues as a sub-category of social problems in general.

While often overlooked, non-academic environmental sociologists are becoming increasingly involved and influential in the research planning for and implementation of mega-projects. They can be found working for government agencies, large engineering and construction firms, consulting companies, architectural firms and local and national NGOs (Payne and Cluett 2002: 526).

How should this considerable evidence of scholarly activity be assessed?

In the late 1970s, Catton and Dunlap undertook a crusade to convert sociologists to their New Ecological Paradigm (NEP)[3] that was meant to cut across the established divisions within sociological theory. This new paradigm was an academic analogue of green thinking in general, advocating an approach that was less 'anthropocentric' (human-oriented) and more 'ecocentric' (humans are only one of many species inhabiting the earth). Buttel (1987: 466) describes their efforts as nurturing a set of 'lofty intentions' wherein environmental sociologists 'sought nothing less than the re-orientation of sociology toward a more holistic perspective that would conceptualise social processes within the context of the biosphere'.

Catton and Dunlap now acknowledge that they failed in this endeavour, but claim that they never fully expected to achieve this kind of disciplinary conversion (Dunlap and Catton

1992/3: 272). More recently, Dunlap (2002b: 21) has distinguished between his expectations in undertaking this project and those of Catton, his collaborator. While Catton 'may have seen the NEP as leading to the development of a truly ecological theory of human societies', he (Dunlap) had more modest ambitions: to sensitise sociologists to the importance of environmental problems in general, and to prod them into recognising our ecosystem dependence.

Both Buttel and Catton and Dunlap have observed that the environmental sociology field faltered during the Reagan era. However, while Buttel pessimistically refers to environmental sociology as having become just 'another sociological specialisation', Catton and Dunlap suggest that the resurgence of interest in environmental issues in the 1990s, especially those which are global in scope, has stimulated renewed interest in environmental sociology in the United States as well as internationally.

In the US, the rising popularity throughout the 1990s of the Environmental Justice Paradigm (EJP) has created new growth opportunities for environmental sociology. In no small way, this has mirrored the rise of the Environmental Justice Movement (EJM) itself. That is, just as the EJM grew by appropriating frames from the labour, civil rights and social justice movements and re-constituting these into a new environmental justice identity (Taylor 2000: 562), the EJM has established a notable presence within American sociology by bridging research into racial and ethnic inequality, urban poverty, occupational health and worker safety and health. The rise to prominence of the EJP is marked by the inclusion on the 2004 American Sociological Association Annual Meeting programme of a plenary session on Love Canal. Among the featured speakers here was Lois Gibbs, the iconic environmental heroine of the Love Canal struggle.

All too rare still are seminal works that could lift environmental sociology into the mainstream of theoretical debate in the broader field of sociology.

One such theoretical soliloquy is Ulrich Beck's book, *Risk Society* (see Chapter 2 for a detailed assessment). Beck, a sociologist of institutions, has approached the subject of environmental risks more from the perspective of a macro-sociology of social change (Lash and Wynne 1992: 8) than from a paradigm that is rooted specifically in environmental sociology. Nevertheless, Beck's argument has been widely noticed and has provoked considerable discussion both within and beyond the confines of environmental sociology. Higgins and Natalier (2004: 80) describe *Risk Society* as a 'seminal' work that has 'become one of the pre-eminent sociological texts; its sweeping scope and impassioned critique of environmental crises have struck a chord with the public and sociologists alike'. It helps, of course, that his concept of 'reflexive modernisation' is shared by Anthony Giddens, the most widely recognised and arguably the most influential sociological theorist in Britain over the last quarter-century.

By contrast, Catton and Dunlap's HEP (Human Exemptionalism Paradigm)/NEP (New Ecological Paradigm) distinction – the most broadly disseminated typology (and methodology) within the area of environmental sociology to date – has failed to generate much excitement outside of this speciality area and its siblings in psychology, political science and environmental education. While Dunlap, in conjunction with the Gallup polling organisation, has exported the HEP/NEP scale internationally, it has not centrally shaped any of the major controversies in environmental social theory in recent years, especially outside North America.

Buttel (2002: 49–50) identifies one 'big idea' in sociology produced by an environmental sociologist (Schnaiberg's [1980] 'treadmill of production') and two other environmental-sociological notions – those of the 'metabolic rift' (Foster 1999) and 'ecological modernisation' (Mol 1997) that seem promising. However, he concedes that neither of these has really had a significant presence outside environmental sociology.

At the end of the day, then, it probably makes sense to embrace Elizabeth Shove's (1994) notion that sociologists can make a positive contribution to the environmental debate by both incorporating and engaging. The former suggests that pockets or niches of environmental research can enrich mainstream sociological theory even if they do not as yet have the capacity to transform the discipline as a whole. The latter recognises that there is much to gain in applying the sociological imagination to the extra-disciplinary study of contemporary environmental issues; for example, through political economy models or via the sociology of science and knowledge. Alas, sociologists far too often end up as 'underlabourers' in this endeavour, being viewed as supporting actors in a cast dominated by natural scientists and environmental policy-makers.

2 Contemporary theoretical approaches to environmental sociology

Environmental sociology, Buttel (2003) observes, has gone through two distinct stages since its emergence in the 1970s as a discrete disciplinary area. In the first stage, the major theoretical task was to identify a key factor (or a closely related set of factors) that created an enduring 'crisis' of environmental degradation and destruction. More recently, there has been a significant shift towards another task: discovering the most effective mechanism of environmental reform or improvement which will help 'chart the way forward to more socially secure and environmentally friendly arrangements' (p. 335).

In this chapter, I will begin by discussing two major approaches to the environment and society that were conceived with the first of these problematics in mind, and then proceed to an overview of two contrasting perspectives, reflexive modernisation and ecological modernisation, which address the second. Next, I recall what has probably been the most enduring, and at times rancorous, debate in the field, the realism–constructionism debate. Finally, I describe some recent attempts to move beyond this dualism and develop a more integrated 'co-constructionist' model of society, nature and the environment.

Two foundational explanations for environmental degradation and destruction

In accounting for the causes of widespread environmental destruction, two primary approaches stand out: the ecological explanation as embodied in Catton and Dunlap's model of 'competing environmental functions', and the political economy explanation as found in Alan Schnaiberg's concepts of the 'societal-environmental dialectic' and the 'treadmill of production'. As Buttel (1987: 471) has noted, both approaches view social structure and social change as being reciprocally related to the biophysical environment but the nature of this relationship is depicted very differently.

Ecological explanation

The ecological explanation for environmental destruction has its roots in the field of 'human ecology' that remained dominant within urban sociology from the 1920s to the 1960s.

This urban ecology model was introduced during the 1920s and 1930s by sociologist Robert Park and colleagues at the University of Chicago. Park was well acquainted with the work of Darwin and his fellow naturalists, drawing on their insights into the interrelation and interdependence of plant and animal species. In his discussion of human ecology, Park (1936 [1952]) begins with an explanation of the 'web of life', citing the familiar nursery rhyme, The House that Jack Built, as the logical prototype of long food chains, each link of which is dependent upon the other. Within the web of life, the active principle is the 'struggle for existence' in which the survivors find their 'niches' in the physical environment and in the division of labour among the different species.

If Park had been primarily interested in the natural environment for its own sake, he might have realised that human intervention in the form of urban development and industrial pollution artificially broke this chain, thereby upsetting the 'biotic balance'. In fact, he did acknowledge that commerce, in 'progressively destroying the isolation upon which the ancient order of nature rested', has intensified the struggle for existence over an ever-widening area of the habitable world. But he believed that such changes had the capacity to give a new and often superior direction to the future course of events forcing adaptation, change and a new equilibrium.

Biological ecology was the primary source from which Park borrowed a series of principles, which he applied to human populations and communities. In doing so, however, he notes that human ecology differs in several important respects from plant and animal ecology. First, humans are not so immediately dependent upon the physical environment, having been emancipated by the division of labour. Second, technology has allowed humans to remake their habitat and their world rather than to be constrained by it. Third, the structure of human communities is more than just the product of biologically determined factors; it is governed by cultural factors, notably an institutional structure rooted in custom and tradition. Human society, then, in contrast to the rest of nature, is organised on two levels: the biotic and the cultural.

This portrait of the nature–society relationship clearly contravenes many of the tenets of Catton and Dunlap's New Ecological Paradigm. It emphasises humans' exceptional characteristics (inventiveness, technical capability) rather than their commonality with other species. It gives priority to the influence of social and cultural factors (communication, division of labour) rather than biophysical, environmental determinants. Finally, it downplays the constraints imposed by nature by celebrating the human capacity to master it.

Park, his colleagues and students (notably McKenzie and Burgess) applied their principles of human ecology to the processes that create and reinforce urban spatial arrangements. They visualised the city as the product of three such processes: (1) concentration and deconcentration; (2) ecological specialisation; and (3) invasion and succession. The building blocks of the city were said to be 'natural areas' (slums, ghettoes, bohemias), the habitats of natural groups that were in accordance with these ecological processes. The city was depicted as a territorially based ecological system in which a constant Darwinian struggle over land use produced a continuous flux and redistribution of the urban population. Nowhere was this more evident than in the 'zone in transition', an area adjacent to the central business district which went from a coveted residential district to a blighted area characterised by low rent tenants, deviant activities and marginal businesses.

Much of the early criticism of human ecology rested not on its failure to explore the interdependence between the human environment and the natural environment but rather in what was perceived as its failure adequately to account for the role of human values in residential choice and movement. In the late 1940s, a sociocultural critique of mainstream human ecology briefly lit up the landscape of American sociology. Firey (1947) used the example of land use in central Boston to demonstrate that symbolism and sentiment were equally, if not more, important than standard ecological principles in accounting for the shape of the city. Similarly, Jonassen (1949) presented the history of settlement and relocation of Norwegian immigrants to the New York City area as evidence that ethnic groups consciously choose a specific type of residential environment on the basis of values which they bring with them as a type of cultural baggage (in this case, the ideal included the sea, a harbour and mountains). Jonassen's research might have been the launching pad for a body of research on the origins of environmental perceptions (see for example Lynch's (1993) article on constructions of nature in Latin America) but the main thrust of his argument was rather to discredit the economic determinism that characterised the orthodox ecology of the day.

While cultural ecology, *per se*, never became dominant, it did force more traditional human ecologists to take greater account of social organisational and cultural variables. This was evident in O. D. Duncan's 1961 POET model (Population-Organisation-Environment-Technology) which was depicted as an 'ecological complex' in which: (1) each element is interrelated with the other three, and (2) a change in one can therefore affect each of the others. The POET model was a trailblazer in providing insight into the complex nature of ecological disruptions even if it failed to give sufficient weight to environmental constraints. For example, in a causal sequence suggested by Dunlap (1993: 722–3), an increase in population (P) can create a pressure for technological change (T) as well as increased urbanisation (O), leading to the creation of more pollution (E). While it was still rooted in orthodox human ecology, nevertheless, Duncan's POET model with its use of the human ecological complex at times 'came close to an embryonic form of environmental sociology' (Buttel and Humphrey 2002).

In all of this, an important issue is whether the notion of an 'ecosystem' should be accepted at face value or merely treated as an analogy. It seems likely that Park and the Chicago School had the latter in mind, adopting the conceptual language of biological ecology because it was the scientific flavour of the day (see Chapter 3). Other social scientists, however, took the ecological metaphor more literally. For example, the noted economist Kenneth Boulding (1950: 6) claimed that he was using the ecosystem concept in its proper sense, and not merely [as] an analogy. Society was, he wrote, 'something like a great pond' filled with 'innumerable "species" of social life, organisations, households, businesses and commodities of all kinds' (1950: 6).[1]

Competing functions of the environment

The ecological basis of environmental destruction is probably best described in Catton and Dunlap's own 'three competing functions of the environment' (see Figure 1). This scheme has been much less widely disseminated than their theory of the 'dominant social paradigm', even though it is, to my mind, more conceptually interesting.

Catton and Dunlap's model specifies three general functions that the environment serves for human beings: supply depot, living space and waste repository. Used as a supply depot, the environment is a source of renewable and non-renewable natural resources (air, water, forests, fossil fuels) that are essential for living. Overuse of these resources results in

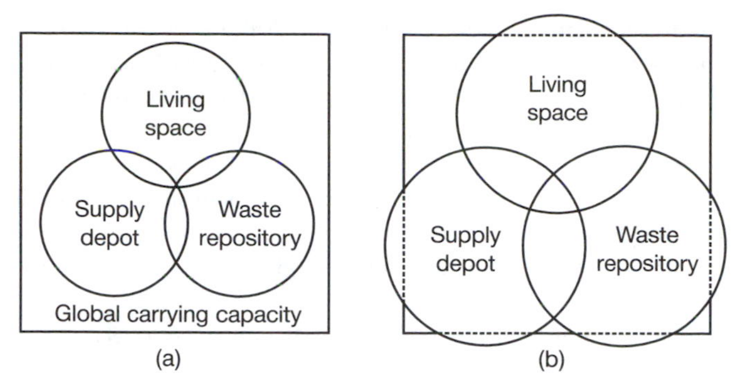

Figure 1 Competing functions of the environment: (a) circa 1900; (b) current situation.
Source: Dunlap 1993

shortages or scarcities. Living space or habitat provides housing, transportation systems and other essentials of daily life. Overuse of this function results in overcrowding, congestion and the destruction of habitats for other species. With the waste repository function, the environment serves as a 'sink' for garbage (rubbish), sewage, industrial pollution and other byproducts. Exceeding the ability of ecosystems to absorb wastes results in health problems from toxic wastes and in ecosystem disruption.

Furthermore, each of these functions competes for space, often impinging upon the others. For example, placing a garbage landfill in a rural location near to a city both makes that site unsuitable as a living space and destroys the ability of the land to function as a supply depot for food. Similarly, urban sprawl reduces the amount of arable land that can be put into production while intensive logging threatens the living space of native (aboriginal) peoples.

In recent years, the overlap, and therefore conflict, among these three competing functions of the environment has grown considerably. Newer problems such as global warming are said to stem from competition among all three functions simultaneously. Furthermore, conflicts between functions at the level of regional ecosystems now have implications for the global environment.

There are several very attractive features to Catton and Dunlap's competing functions of the environment model. First and foremost, it extends human ecology beyond an exclusive concern with living space – the central focus of urban ecology – to the environmentally relevant functions of supply and waste disposal. In addition, it incorporates a time dimension: both the absolute size and the area of overlap of these functions are said to have increased since the year 1900.

At the same time, there are problems with the model. As is the case with the urban ecology of Park and the Chicago School, there is no evidence of a human hand here. It says nothing about the social actions involved in these functions and how they are implicated in the overuse and abuse of environmental resources. Above all, there is no provision for changing either values or power relationships. The former is especially puzzling, since one would have thought that Catton and Dunlap would have attempted to link their ecological model to the new human ecology as emphasised in the HEP/NEP contrast. Finally, one cannot help comparing the longitudinal features of the Catton–Dunlap model to Beck's

(1992) depiction of the transformation from an industrial to an industrial risk society. Both models recognise some of the same features: the increasing globalisation of environmental dangers, the rising prominence of output or waste-related elements as opposed to input or production-related ones. However, Beck's model is ultimately more exciting because it centrally incorporates the process of social definition. Beck's (1992: 24) criticism of environmental risk assessment, i.e. that 'it runs the risk of atrophying into a discussion of nature without people, without asking about matters of social and cultural significance', is equally applicable to Catton and Dunlap's competing functions of the environment.

Political economy explanation: the 'societal–environmental dialectic' and the 'treadmill of production'

Within environmental sociology, probably the most influential explanation of the relationship between capitalism, the state and the environment can be found in Alan Schnaiberg's book, *The Environment: From Surplus to Scarcity* (1980). Drawing on strands of both Marxist political economy and neo-Weberian sociology, Schnaiberg outlines the nature and genesis of the contradictory relations between economic expansion and environmental disruption.

Schnaiberg has depicted the political economy of environmental problems and policies as being organised within the structure of modern industrial society, which he labels the treadmill of production. This refers to the inherent need of an economic system to continually yield a profit by creating consumer demand for new products, even where this means expanding the ecosystem to the point where it exceeds its physical limits to growth or its 'carrying capacity'. One particularly important tool in fuelling this demand is advertising, which convinces people to buy new products as much for reasons of lifestyle enhancement as for practical considerations.

Schnaiberg portrays the treadmill of production as a complex self-reinforcing mechanism whereby politicians respond to the environmental fall-out created by capital intensive economic growth by mandating policies that encourage yet further expansion. For example, resource shortages are handled not by reducing consumption or adopting a more modest lifestyle but by opening up new areas to exploitation.

Schnaiberg detects a *dialectic tension* that arises in advanced industrial societies as a consequence of the conflict between the treadmill of production and demands for environmental protection. He describes this as a clash between 'use values'; for example, the value of preserving existing unique species of plants and animals, and 'exchange values' which characterise the industrial use of natural resources. As environmental protection has emerged as a significant item on the policy agendas of governments, the state must increasingly balance its dual role as a facilitator of capital accumulation and economic growth and its role as environmental regulator and champion.

From time to time, the state finds it necessary to engage in a limited degree of environmental intervention in order to stop natural resources from being exploited with abandon and to enhance its legitimacy with the public. For example, in the progressive era of American politics in the late nineteenth and early twentieth centuries, the US government responded to uncontrolled logging, mining and hunting on wilderness lands by expanding its jurisdiction over the environment. Especially under the presidency of Theodore ('Teddy') Roosevelt, it created national forests, parks and wildlife sanctuaries, set limits and rules for the use of public lands and restricted the hunting of endangered species. It did so, however, as much out of a desire to increase industrial efficiency (Hays 1959), regulate competition and ensure a steady supply of resources (Modavi 1991) as it did from any sense of moral outrage. Similarly, the sudden emergence of toxic waste as a premier media issue in the early 1980s led to Congressional efforts in the United States to pass a new 'Superfund' law that would give the government statutory authority and the fiscal mechanisms to undertake clean-up operations without first having legally to identify the responsible parties. This was, Szasz (1994: 65) notes, not simply a matter of lawmakers addressing a newly recognised social need, but instead 'one of those quintessential "time to make a new law" moments so characteristic of the American legislative process'. Nevertheless, most governments remain wary of running the risk of slowing down the drive towards economic expansion or decelerating the treadmill of production (Novek and Kampen 1992). Caught in a contradictory position as both promoter of economic development and as environmental regulator, governments often engage in a process of 'environmental managerialism' (Redclift 1986), in which they attempt to legislate a limited degree of protection sufficient to deflect criticism but not significant enough to derail the engine of growth. By enacting environmental policies and procedures that are complex, ambiguous and open to exploitation by the forces of capital production and accumulation (Modavi 1991: 270) the state reaffirms its commitment to strategies for promoting economic development.

Other more stridently left-wing critiques have been even more unsparing in linking the dynamics of capitalist development to the rise in environmental destruction. David Harvey (1974), the Marxist geographer, accuses capitalist supremos of deliberately creating resource scarcities in order that prices may be kept high. Faber and O'Connor (1993) charge that the goal of capital restructuring in the 1980s and 1990s, which included geographical relocation, plant closures and downsizing, is to increase the exploitation of both the workers and nature; for example, by reducing spending on pollution control equipment. Cable and Cable refuse to rule out the possibility of insurrection in the United States if the grievances of grassroots environmental groups continue to be ignored by capitalist economic institutions (1995: 121). Schnaiberg himself (2002: 33) has complained that the central tenets of the treadmill have not found their way into the environmental sociological literature in any significant way because they are too 'radical'. That is, if the treadmill was indeed operating as he describes, then it can only be altered by a major and sustained political mobilisation, something that would be sharply resisted by politicians, government agencies and corporate America.

Subsequently, the 'treadmill of production group'[2] has addressed the application of the treadmill of production to a Third World context. Ignoring the negative environmental impacts that the treadmill has produced in less developed regions, the leaders of Southern nations, in concert with the governments and corporations of the North, have sought to reproduce industrialisation as experienced by the First World. The primary mechanism for achieving this is the transfer of modern Western industrial techniques from North to South (Schnaiberg and Gould 1994: 167). However, as Redclift (1984) and others have noted, this transplant has become largely unsuccessful both in economic and environmental terms. Dependency on global markets has made economic development a risky venture for many Third World nations, especially where these markets can easily be decimated by the appearance of new, low-cost alternatives elsewhere in the world. Furthermore, development schemes require an expensive infrastructure of roads, hydroelectric power dams, airports, and so on, which must be paid for by borrowing heavily from Northern financial institutions. Such projects often fail to produce the expected level of economic growth while at the same time causing massive ecological damage in the form of flooding, rainforest destruction, soil erosion and pollution.

The treadmill of production explanation has the advantage of locating present environmental problems in the inequities of humanly constructed political and economic systems rather than the abstract conflict of functions preferred by human ecologists. This brings it closer to the orbit of mainstream sociological theory than the more idiosyncratic approach advocated by Catton and Dunlap. At the same time, as Buttel (2004: 323) has observed, the concept of the treadmill is unique insofar as it is based in sociological reasoning but, at the same time, features a key or penultimate dependent variable – environmental destruction – that is biophysical. In Buttel's judgement, this makes it 'the single most important sociological concept and theory to have emerged within North American environmental sociology'.

As Schnaiberg himself has recognised, the treadmill of production has not achieved the paradigmatic status within environmental sociology that he would have liked. Buttel offers several possible reasons for this. First, political economy, especially that with a neo-Marxist hue, has been somewhat overshadowed in recent decades by other theoretical flavours, notably postmodernism and cultural sociology. Second, treadmill theory has remained somewhat static, wedded to a manufacturing economy in a neo-liberal era in which Western economies seem to have shifted towards new information technologies, financial services and entertainment. Another reason may be simply that the notion of the treadmill is no longer very new or, in spite of what Schnaiberg believes, very controversial. To actually shut down the treadmill, of course, *would* be quite radical, but as an analysis of industrial and consumer society the model seems rather obvious, something that might not have been the case thirty years ago.

Two normative theories of modernism and environmental improvement

In considering mechanisms of environmental improvement, Buttel (2003) proposes four potential channels: environmental activism/movements (he judges this to be the most fundamental and promising), state environmental regulation, ecological modernisation, and international environmental governance.

Theoretically speaking, two recent models stand out here, both normatively charged, late modernist prescriptions emanating from Germany and Holland. These are Beck's 'risk society thesis' and Mol and Spaargaren's 'ecological modernisation' (EM) theory. The two approaches have often been pitted against one another, insofar as the latter is intended to transform economy–ecology contradictions into win–win situations, while the former claims that our efforts to reform industrial society in the face of an apocalyptic eco-societal crisis are Herculean, if not futile (Blowers 1997; Desfor and Keil 2004: 62). At the same time, the two approaches share an important commonality: the expectation that an 'environmental state' will eventually emerge, where environmental protection is a basic responsibility (Fisher 2003: 9–10).

Risk society thesis

As we begin the new millennium, probably the most influential attempt to update modernism has been Ulrich Beck's 'risk society thesis'. In comparison to EM theory, Beck is openly critical of modernity and its attendant risks. Nevertheless, he concludes that modernity ultimately has the capacity to solve the problems it produces (Barry 1999: 152).

Beck's thesis starts with the premise that Western nations have moved from an 'industrial' or 'class' society in which the central issue is how socially produced wealth can be distributed in a socially unequal way while at the same time minimising negative side effects (poverty, hunger) to the paradigm of a 'risk society' in which the risks and hazards produced as part of modernisation, notably pollution, must be prevented, minimised, dramatised or channelled. In the case of the latter, risk is said to be much more evenly distributed than was formerly the case. As Beck phrases it, 'hunger is hierarchical, smog is democratic'. Nevertheless, both the former 'wealth distributing society' and the emergent 'risk distributing society' contain inequalities and these overlap in areas such as the industrial centres of the Third World.

Contemporary risks are set apart from those of the past through their origins, scope and effect and the difficulties of identification (see Higgins and Natalier 2004: 78–9). Risk attached to events such as chemical spills and radiation poisoning are more than the unfortunate byproducts of industrialism and capitalism. Rather, they are a testament to the failure of social institutions, most notably science, to control new technologies. Such risks transcend both space and time, extending well beyond the geographic source, and temporally, beyond the present generation. The 1986 Chernobyl nuclear accident in the Ukraine is a dramatic illustration of this. Due to the 'boomerang effect', risks that are exported abroad, notably to the nations of the South, inevitably come back to haunt us. Finally, risks today are said by Beck to be largely invisible to lay people, identifiable only through sophisticated scientific instrumentation.

One important feature of the risk society is the way in which the past monopoly of the sciences on rationality has been broken. Paradoxically, science becomes 'more and more *necessary*, but at the same time, *less and less sufficient* for the socially binding definition of truth' (Beck 1992: 156). Beck contrasts the rigid 'scientific rationality' that prevailed for most of the twentieth century with a new 'social rationality' that is rooted in a critique of progress. Under pressure from an increasingly edgy public, new forms of 'alternative' and 'advocacy' science come into being and force an internal critique. This 'scientisation of protest against science' produces a fresh variety of new public-oriented scientific experts who pioneer new fields of activity and application (e.g. conservation biology). In a similar fashion, monopolies on political action are said to be coming apart, thus opening up political decision-making to the process of collective action. One example of this is the entry of the 'greens' into parliamentary politics in Germany in the 1980s.

Finally, the dynamic of reflexive modernisation leads to a greater individualisation. Unbound from the strictures of traditional, pre-modern societies, the new urban citizens of the industrial revolution were supposed to reach new levels of creativity and self-actualisation. However, this did not happen, largely because a new constraint – the 'culture of scientism' – invaded every part of our lives from risk construction to sexual behaviour. Now there is a chance for individuals once again to break free and choose their own lifestyles, subcultures, social ties and identities (p. 131). Each of us, Beck believes, is obliged to reflect upon our personal experiences and make our own decisions about how we wish to live (Irwin 2001: 56). Yet, ironically, just as the individualised private existence finally becomes possible we are confronted with risk conflicts, which by their origin and design, resist any individual treatment. 'Global environmental problems' such as the greenhouse effect and the thinning of the ozone layer are key illustrations of this. Thus the 'reflexive scientisation' in which scientific decision-making, especially that related to risk, is opened up to social rationality is vital to the reclamation of individual autonomy. Democracy should not, he insists, 'end at the laboratory door'.

While Beck's analysis is fresh and powerfully presented, it is not without its problems.

As Lidskog (1993) has pointed out in his review of *Risk Society*, Beck contradicts himself by arguing that the planet is in increasing peril due to an escalation of objectively certifiable global risks and, at the same time, insisting that risks are entirely socially constructed and therefore do not exist beyond our perception of them. Blühdorn (2000: 86) too hones in on this inconsistency, pointing out that Beck 'seems to be undecided whether ecological risks have to be conceptualised as objective empirical realities or as subjective perceptions and social constructions'. Indeed, if you were to question Beck's assertion that the scope and effect of 'real' risk has sharply increased in late modernity, then this would have serious implications for the efficacy of his entire 'risk society' thesis.

More generally, Beck's inconsistency on this point reflects a long-standing tension in environmental sociology between the role of the sociological analyst and that of the environmental activist. Catton and Dunlap's HEP/NEP dichotomy is the epitome of this but it runs through much of the rest of the literature as well, surfacing, for example, in the 'critical realist' approach of Benton, Dickens, Martell and other British sociological thinkers who seek to put nature back into the nature–society relationship. In Beck's risk society thesis, descriptive and prescriptive dimensions continually interweave. Indeed, Beck appears to be actively promoting a distinctive vision of an 'ecologically rational' or 'ecologically enlightened' society (Barry 1999: 153).

Beck's response to this criticism is frustrating. He sees no essential contradiction between depicting a world in which risk is pervasive and possibly apocalyptic while observing that such risks are 'particularly open to social definition and construction' (1992: 23).

Even more fundamentally, Beck conflates and confuses the meaning of risks and hazards. On the one hand, he defines risk as 'a systematic way of dealing with hazards and insecurities introduced by modernization itself' (1992: 21). While citizens in a pre-industrial society were no strangers to hazards – famines, plagues, natural disasters – no notion of risk was to be found, because hazards or dangers were experienced as pre-given, usually as punishments from the gods (Elliott 2002: 295). Yet, as we have seen, Beck's theory of social change rests on the assertion that risks in a globalised society are both more extensive and more democratically distributed than was true before. Furthermore, it implies that risks such as those related to nuclear power plants and runaway biotechnology are hazardous *in and of themselves* rather than constituting new ways of defining and coping with these hazards (Sutton 2004: 121).

Beck has also attracted considerable critical heat for his assertion that class-based rancour over the distribution of goods has fallen off in favour of new and shifting patterns of coalition and division. Increasingly, he ventures, it is not unusual to observe situations where workers in environmentally polluting industries join together with management in opposition to 'victims' from competing sectors of the economy such as fisheries and tourism. In some cases, alliances may even emerge between those once seriously in conflict with one another. For example, in New Mexico and Montana, ranchers and green organisations such as the Sierra Club have recently put aside their historic differences to jointly battle against the common threat of proliferating oil and gas wells (Carlton 2005). This interpretation is flawed, however, in that powerless economic actors are frequently compelled to support polluting technologies and policies in order to survive. Citing the case of Australian broadacre farmers who have come to accept chemical-dependent styles of agriculture as rational approaches to environmental management, Lockie (1997) notes that it is possible to be both a 'victim' and a 'perpetrator' at one and the same time. That is, the farmer as perpetrator contributes to global pollution through engaging in chemical-intensive farming practices even as the farmer as victim is exposed to toxic materials that may be the source of chemically-induced illness, ranging from headaches to cancer.

Critics of the risk society thesis have accused Beck of being unacceptably vague about the details of political and scientific decision-making in the reflexive phase of modernity that he sees as imminent. Seippel (2002: 215–6) implies that Beck's vision of politics in a 'civil society' is naïve and utopian. Why should we expect the political jockeying and dealing that are characteristic of traditional politics suddenly to disappear overnight? Indeed, in blurring the boundaries between conventional politics and civil society, we may even risk opening the latter up to undemocratic interests, values or modes of action. Furthermore, Beck overstates the potential for ecological rationality here, ignoring the 'cultural embeddedness' of social interaction. That is to say, there is little reason to expect that a society obsessed with celebrities and shopping will suddenly change direction and start making choices solely on the basis of new, post-materialist values. In short, as enlightening as it may seem, the risk society thesis ultimately constitutes a 'mythical discourse' (Alexander and Smith 1996, cited in Seippel 2002: 215).

Ecological modernisation

By ecological modernisation, Spaargaren and Mol mean an ecological switch of the industrialisation process in a direction that takes into account the maintenance of the existing sustenance base (1992: 334). Cast in the spirit of the Bruntland Report, ecological modernisation, like sustainable development, 'indicates the possibility of overcoming the environmental crisis without leaving the path of modernization'. The model is based on the work of the German writer, Huber (1982; 1985) who analyses ecological modernisation as a historical phase of modern society. In Huber's scheme, an industrial society develops in three phases: (1) the industrial breakthrough; (2) the construction of industrial society; and (3) the ecological switchover of the industrial system through the process of 'superindustrialisation'. What makes this latter phase possible is a new technology: the invention and diffusion of microchip technology.

Ecological modernisation rejects the 'small is beautiful' ideology inspired by Schumacher (1974) in favour of large-scale restructuring of production–consumption cycles to be accomplished through the use of new, sophisticated, clean technologies (Spaargaren and Mol 1992a: 340). Unlike sustainable development, there is no attempt to address problems of the less developed countries of the Third World. Rather, the theory focuses on the economies of Western European nations which are to be 'ecologised' through the substitution of microelectronics, gene technology and other 'clean' production processes for the older, 'end-of-pipe' technologies associated with the chemical and manufacturing industries. In contrast to Schnaiberg's 'treadmill of production' perspective, capitalist relations of production, operating as a treadmill in the ongoing process of economic growth, are treated as largely irrelevant (Spaargaren and Mol 1992: 340–1)

According to Udo Simonis (1989), a German environmental policy analyst, the ecological modernisation of industrial society contains three main strategic elements: a far-reaching conversion of the economy to harmonise it with ecological principles, a reorientation of environmental policy to the 'prevention principle' (seeking a better balance between stopping pollution before it happens and cleaning it up later on) and an ecological reorientation of environmental policy, especially by substituting statistical probability for 'prove-beyond-a-doubt' causality in legal suits against polluters. Unfortunately, little is said about the social and political barriers that are likely to be faced in trying to implement these strategies, especially in countries other than Germany and the Netherlands where the environment is a major priority.

Ecological modernisation thinkers are to be commended for attempting to stake out a reasoned position between 'catastrophic' environmentalists who preach that nothing less than de-industrialisation would suffice in saving the Earth from an ecological Armageddon and capital apologists who prefer a business-as-usual approach (Sutton 2004: 146). Alas, the ecological modernisation perspective is hobbled by an unflappable sense of technological optimism.[3] All that is needed, they suggest, is to fast-forward from the polluting industrial society of the past to the new super-industrialised era of the future. Yet, the silicon chip revolution, which is the basis of this super-industrialisation, is by no means environmentally neutral as the theory of ecological modernisation suggests (see Mahon 1985). Furthermore, it is worth remembering that nuclear power was also touted as a 'clean' technology until its more undesirable features became known.

As a sociological explanation, the theory of ecological modernisation is as much prescriptive as analytic. Spaargaren and Mol, for example, initially said little about the power relations that characterise environmental processes, assuming that somehow good sense must automatically triumph. Yet, as Gould *et al.* (1993: 231) have argued, sustainability, the guiding concept behind ecological modernisation, is as much a political–economic dimension as an ecological one: what can be sustained is only what political and social forces in a particular historical alignment define as acceptable. Recognition of this is far more evident in Beck's concept of a risk-distributing society than in the ecological modernisation which Mol and Spaargaren see as imminent.

More recently, Mol and Spaargaren have offered up a revisionist version of ecological modernisation theory. The initial debates of the early 1980s, they caution, 'should be understood as an overreaction directed at the dominant schools of thought in environmental sociology and the environmental debate in the late 1970s and early 1980s' (2000: 18–19). In particular, ecological modernisation theory, they insist, was originally meant to challenge the notion put forward by both neo-Marxists and counter-productivity thinkers such as Rudolph Bahro and Barry Commoner that the modernisation project was in its death throes; that the widespread environmental and ecological deterioration of the time was *prima facie* evidence of this; and that things could be salvaged only by fundamentally recognising the core institutions of modern society.

Today, Mol and Spaargaren claim, these initial debates have become less relevant. Significantly, capitalism itself has evolved in a greener direction. For example, market-based instruments such as tradeable pollution credits have displaced previous strategies that emphasised heavy-handed state regulation and enforcement. Furthermore, ecological modernisation theorists themselves have incorporated critical comments from the earlier debate, reforming and refining their analysis of social change. For example, they now claim to present a more nuanced position regarding capitalism, interpreting it 'neither as an essential precondition for, nor as the key obstruction to, stringent and radical environmental reform' (2000: 23).

Whereas the initial debate was frequently waged with neo-Marxists, now Mol and Spaargaren confide that they are making 'new theoretical alliances' (2000: 25) with them against their common foes – postmodernists and social constructionists. Political economists and ecological modernisationists, they argue, converge and agree in their criticism against strong social constructionism and in their view that environmental problems have a 'real' existence. Both can be considered as branches of the modernist project, assuming a firm stance against postmodern analyses of environmental problems and solutions (Mol and Spaargaren 2002: 35).

Mol and Spaargaren say they are irritated that outdated positions and criticisms from the 1970s and 1980s keep re-appearing with some regularity. For example, proponents of the New Environmental Paradigm continually threaten to go overboard, replacing sociology's former disregard of nature 'with some form of present-day biologism or ecologism' (2002: 27). Even more problematic, they assert, are those postmodern authors, most notably Blühdorn (2000), who depict the ecological crisis as merely another 'grand narrative' to be deconstructed; and ecological rationality as 'nothing more than power, politics and big money'. This same virulent strain is evident in the views of 'hard' or 'strict' social constructionists. Even Maarten Hajer (1995), whose case history of ecological modernisation as it is manifested in the politics of acid rain has been widely praised, is evidently considered to be suspect insofar as he 'seems to end up taking a position which is not too far away from where postmodernity would feel comfortable' (2002: 30). Finally, radical eco-centrists are dismissed because they criticise ecological modernisation for advocating a watered-down form of environmentalism which assumes that the crisis of the earth can be resolved by modifying attitudes, laws, government policies, corporate behaviour and personal lifestyles rather than by demanding fundamental structural change. Being in the camp of the radical ecologists, they warn, is 'about being a pessimist by nature' (2002: 33).

Despite their apparent *rapprochement* with the Schnaiberg school of political economy, Mol and Spaargaren still seem to place their faith in 'responsible capitalism' and the primacy of the market. For example, in his empirical research into the ecological modernisation of production in the Dutch chemical industry, evidently a notorious polluter in the past, Mol (1997) finds nothing but good news. Reacting to consumer pressure, Dutch chemical companies have initiated a spate of green measures, from the introduction of new technologies (low organic solvent paints) to new corporate instruments such as annual environmental reports, environmental audits and environmental certification systems. Together, he says, this represents 'a process of radical modernization' that has undercut any misguided 1970s and 1980s style demands for the dismantling of chemical production or even a shift to 'soft chemistry' (e.g. 'natural paints', which have failed to capture more than a one per cent share of the market in European countries). The institutions of modernity, Mol concludes, are by no means fading away; no massive movement away from a 'chemicalised' lifestyle can be identified and the erosion of trust in the scientific foundations of the chemical industry that might be inferred from Beck's risk society thesis is more or less absent.

Contributors to the treadmill of production perspective, however, are considerably less enamoured of ecological modernisation theory than vice versa. In the definitive statement on this in a 2002 collection of articles entitled *The Environmental State Under Pressure*, Schnaiberg and his associates deny that the best hope for solving environmental problems is to embrace new technologies. In America, at least, environmental policy-making continues to be written within an economic framework and the green movement has failed to become a major political force. This is evident, they argue, in both industry evasion and dilution of recycling controls, and in the failure of the highly touted President's Council on Sustainable Development during the Clinton administration (1993 to 1999). Such cases fundamentally challenge the core postulates of ecological modernisation theory.

Why do the treadmill analysts differ so broadly from the ecological modernisationists? Schnaiberg suggests, rather diplomatically, that it has to do with a difference in sampling approaches. That is, ecological modernisation (EM) theorists examine 'cutting edge' corporate innovations or 'best practice' industries and assume that these changes will eventually diffuse widely. Treadmill theorists are sceptical, observing that the EM successes heralded by Mol and his colleagues may simply represent a 'creaming' of a programme of ecological incorporation into production practices (Schnaiberg *et al.* 2002: 29). In short, EM theorists are said to be naïve for claiming that greener production practices in arenas such as the Dutch chemical industry constitute a powerful 'third force' and part of a trajectory toward a future characterised by sustainability. Rather, firms that make ecological improvements do so either under direct pressure from state regulation or social movement action. Alternatively, these improvements are not real, having been achieved only through 'creative accounting' or misreporting (p. 29).

To be fair, ecological modernisation theory has become 'an important lens through which changing economy–ecology relationships of industrial societies can be viewed' (Desfor and Keil 2004: 55). This is especially true for the policy-making arena where it has been widely embraced. Nevertheless, as Davidson and Frickel point out:

> For every empirical study supportive of the potential for ecological modernization, there are now a number of empirical analyses that raise numerous caveats regarding the propensity for industry actors to undergo the 'greening' process of their own accord, particularly when we move beyond the advanced countries of Western Europe.
>
> (2004: 477)

At the end of the day, then, whether you regard environmental modernisation as visionary or deluded is ultimately a measure of your degree of faith in gradualism as against the necessity of more radical solutions. As Eckersley (2004: 74) has cautioned, ecological modernisation may well be able to promote greener growth through technological innovation, but eventually it risks being unmasked as 'an ideology free zone'. The more serious ecological problems persist, the more likely this is to occur.

A major controversy: the realism vs. constructionism debate

As Freudenburg (2000: 103) has noted, 'more than any other subject in the discipline in environmental sociology, social construction[ism] has found fertile ground as well as fierce criticism'. Some analysts have accorded social constructionism prime paradigmatic status, situating it at the very core of environmental theory. The idea that the environment is *socially constructed*, Lockie (2004: 29) notes, is 'perhaps one of the most fundamental concepts within environmental sociology'. Others have rejected this claim to being a full-blown, coherent theory as being 'exaggerated', arguing that, at best, it should be seen in more modest terms as 'a set of concepts and methodological conventions' (Buttel et *al.* 2002: 25).

Even less charitable critics from other disciplines have depicted the social constructionist as a sort of Darth Vader, perverting the force of sociological understanding and ignoring the 'reality' of the environmental crisis. For example, the noted conservation biologist Michael Soulé has condemned social constructionism as an academic 'fad' whose rhetoric 'justifies further degradation of wildlands for the sake of economic development' and whose relativism 'can be just as destructive to nature as bulldozers and chainsaws' (Soulé and Lease 1995: xv, cited in Smith 1999: 362). In the same vein, environmental ethicist Eileen Crist (2004: 16) places constructionist analyses of nature in 'the comfort zone of zestless agnosticism and non-committal meta-discourse' where it is foolishly 'striving to interpret the world at an hour that is pressingly calling for us to change it'.

After having raged for a decade the 'constructionist–realist' debate has recently begun to settle, with proponents and opponents alike acknowledging that these sometimes sharp exchanges have become repetitive and counterproductive. Nonetheless, it is worth spending some time recalling why social constructionism first emerged as a way of dealing with environmental matters; what forms it assumed; why it has generated so much critical heat; and how it might continue to make a useful contribution.

Social constructionists are routinely pilloried for allegedly denying that the Earth is under siege from a host of environmental hazards ranging from nuclear power leaks to global warming. This is a grave misrepresentation. Only a 'false reductionism', Wynne (2002: 472) says, can construe constructionist accounts as claiming that environmental risks do not exist or that natural reality plays no identifiable role in producing knowledge about these risks. What constructionists are actually saying is that we need to look more closely at the social, political and cultural processes by which certain environmental conditions are defined as unacceptably risky, and therefore, contributory to the creation of a perceived 'state of crisis'. As Thompson (1991) has noted, environmental debates reflect the existence not just of an absence of certainty but rather of *contradictory certainties*: several divergent and mutually irreconcilable sets of convictions both about the difficulties we face and the available solutions.

Not surprisingly, this multiple and contradictory uncertainties argument irritates constructionist opponents who see it as lending tacit support to those who would deny the existence of environmental problems for their own selfish economic or political reasons. For example, Williams (1998: 486) cites the actions of the Western Fuels Association, a US industry trade group, in reprinting and distributing articles that express uncertainty about specific scientific issues related to global warming, as evidence that powerful social interest groups will exploit any weakness created by constructionist expressions of scientific uncertainty. By contrast, a more 'reflexive realist' view asserts that 'the physical destruction of the environment can be empirically measured and scientifically monitored, thus avoiding an extreme form of naïve constructionism' (Picou and Gill 2000: 145).

Furthermore, critics charge that the conflicting uncertainties approach that has been adopted by constructionists privileges a contingent of 'rogue' scientists over the 'responsible' majority. For example, it is alleged that there is currently a unanimous scientific consensus that the Earth is heating up and that this global climate shift is primarily due to humanly produced greenhouse gas emissions (see Oreskes 2004). The small handful of scientists who dissent from this view, it is argued, are not legitimate because they are firmly 'in the pocket' of various corporations, state officials and anti-climate change interest groups who simply do not want to make the costly policy changes that would be required to comply with international accords such as the Kyoto Protocol (Buttel *et al.* 2002: 23). Indeed, for opponents of Kyoto, the vital strategic task is allegedly to keep the public believing that there is no consensus about global warming in the scientific community. And here, it is said, is where constructionists näively betray the environmental cause by encouraging this 'fiction'.

In reply, constructionists argue that bestowing *absolute* certainty solely on the basis of a scientific head count is surely perilous. After all, scientific consensus once dictated that the Earth was flat and that the primary source of disease was 'vapours'. In the case of global warming, the debate is by no means closed. One survey by Dennis Bray and Hans von Storch (2005) of the German Institute for Coastal Research found that as many as a quarter of the 500 international climate researchers who responded to their survey still were not fully convinced that human activity is responsible for the recent rise in global temperatures.

In fact, health and environmental threats do not always follow a unidirectional path.

One recent example of this is the so-called 'killer obesity' crisis. In the wake of a high profile article in the *Journal of the American Medical Association* (9 March 2004) reporting that 'obesity' had caused 400,000 deaths in 2000, up 33 per cent from a decade earlier, poor eating habits were confirmed as a major preventable killer. A year later, the study's authors, the US Centers for Disease Control, corrected these figures, downsizing the number of obesity deaths to 26,000 and revealing that 86,000 moderately overweight Americans were actually found to have *lived longer* than people of normal weight (Henninger 2005). By mid-2005, the pendulum had begun to swing back, as indicated by an article in the *Scientific American* entitled 'Obesity: an overblown epidemic?' (Gibbs 2005).

This does not mean that we should not worry about the alarming incidence of obesity rates, especially among children. Nor should we relax our concerns about the possibility of the polar ice caps melting in the foreseeable future. What it does mean is that it is not wise to allow a *discussable issue* to become an *evident crisis*, especially where the evidence is open to multiple interpretations.

As I have noted in Chapter 1, the first generation of environmental sociologists more or less uncritically accepted the existence of an environmental crisis brought on by unchecked population growth, over-production and the adoption of dangerous new technologies. In particular, Dunlap and Catton's NEP, that 'provided the template for modern environmental sociology' (Buttel 2000: 19) is basically an analogue for the ecocentric claims of radical ecologists that nature must be placed 'at the centre of moral concern, politics and scientific study' (Sutton 2004: 78). Buttel *et al.* (2002: 22) point out that 'prior to the late 1980s, a sizable share of the North American environmental sociology community saw its mission as being to bring the ecological sciences and their insights to the attention of the larger sociological community'. It was dominated by an environmental realism that was 'driven by the impulse of "saving the Earth", pointing to the ongoing environmental destruction and a future global catastrophe' (Lidskog 2001: 120). In this context, constructionism has been labelled as a 'spoiler'.

While not denying the validity of concern over pollution, energy shortages and runaway technology, social constructionists nonetheless insist that the central task ahead for environmental sociologists is *not* to document these problems but to demonstrate that they are the products of a dynamic social process of definition, negotiation and legitimation. As Yearley (1992: 186) observes, demonstrating that an environmental problem has been socially constructed is not to undermine and debunk it, since 'both valid and invalid social problem claims have to be constructed'. Along similar lines, Dryzek notes:

> Just because something is socially interpreted does not mean it is unreal. Pollution does cause illness, species do become extinct, ecosystems cannot absorb stress indefinitely, tropical forests are disappearing. But people can make very different things of these phenomena and – especially – their interconnections, providing grist for political dispute.
>
> (2005: 12)

In short, social constructionism does not deny the considerable powers of nature. Rather, it asserts that the magnitude and manner of this impact is open to human construction.

Furthermore, constructionists maintain that the rank ordering of environmental problem claims by social actors does not always correspond to actual need; rather, it reflects the political nature of agenda setting. Thus, Yearley concludes that:

> There are good grounds for believing that the topics that rise to the top of the public's attention are not those where the reality of the problem is most well documented or where the real impacts are the greatest, but those where the agents that propel issues into the public consciousness have worked the most effectively.
>
> (2002b: 276)

Critics of constructionism have objected to this latter statement. How, they ask, is it possible to determine *actual* need if all things are ultimately nothing but social constructions? Indeed, this would seem to require the sociological analyst to abandon any serious attempt at maintaining an agnostic stance and jump into the fray in order to sort out which claims are convincing and which ones are not.

Furthermore, how does one adjudicate among the competing interpretations of environmental problems?

A failure to do so, Brulle (1998: 138–9) argues, means that 'the social constructivist approach fundamentally undermines the legitimacy of the arguments that environmental problems are real and legitimate and thus deserve our attention for their resolution'.

Williams (1998) agrees and criticises the constructionist position as being inadequate 'because it leads to a relativizing perspective where no claim to reality is privileged over any other' (p. 478). He explicitly recommends that researchers select between competing constructions of environmental problems since the consequences of not doing so are potentially profound.

Hold on!, retort constructionists. Intervention of this type is inherently risky because most sociologists have little formal training in the environmental sciences and are therefore not very well qualified to evaluate the truth or power of environmental claims, especially those that are global in scope (Buttel and Taylor 1992; 1994).

Additionally, social constructionists have also been accused of engaging in the strategy of 'ontological gerrymandering' (Woolgar and Pawluch 1985). By this is meant that constructionist authors continue arbitrarily to identify problematic conditions or behaviour worthy of study at the same time as relativising the definitions and claims made about them. Typically, a condition is treated as objectively real and constant over time while the social evaluation of this condition as problematic or not varies from era to era. This is internally inconsistent, Woolgar and Pawluch argue, since it distinguishes between a set of fixed conditions as identified by the analyst and a set of changing, contextual conditions as proposed by environmentalists and other claims-makers.

In a clever turn of the phrase, Kidner (2000: 343) charges that Hannigan (1995) 'has his epistemological cake and eats it'. By this he means that I and other social constructionists are allegedly inconsistent, claiming at one and the same time that 'environmental problems and solutions are end products of a dynamic social process of definition, negotiation and legitimation' (Hannigan 1995: 24) even as they acknowledge that we face some real and very worrisome global environmental threats.

Social constructionists have responded to these charges in several ways.

First of all, they insist that any claim can be evaluated on the basis of hard evidence such as statistics or public opinion polls, even if these are in themselves social constructions (Best 1989: 247). In particular, the researcher is encouraged to consider the historical context within which the claim has been formulated in order to explain the emergence and assess its validity (Rafter 1992). Agnosticism does not mean that we must automatically accord all claims equal weight. For example, we might reasonably doubt the widely publicised media claims by the Raelians, a flying saucer cult, that they have successfully cloned several humans in their laboratory. On the other hand, a warning from prominent public health officials that tens of millions of urban residents could become clinically ill during a potential outbreak of a global influenza pandemic next winter would carry more authority, even if it is not a certainty.

In making the case for a social constructionist framework for environmental sociology, its practitioners have cited several key advantages.

First of all, social constructionism is said to be more congruent with the existing canon of sociological theorising than are other approaches. Thus, Greider and Garkovitch (1994) argue that the role of the environmental sociologist should lie not in a quest for some elusive new model that causally links ecosystem breakdown with social variables (see Catton 1994) but in a return to classic sociological questions of perception and power. In this context, biophysical changes in the environment are meaningful only insofar as groups affected by these changes come to acknowledge them through a self-redefinition. For example, in addressing the political conflict in the American Northwest over the spotted owl, the key question for the sociological analysts should not be the number of owls but the way in which the fluctuating power of the different social actors or claims-makers – loggers, rural businesses, international logging companies, environmentalists – shapes the definition of the situation (Greider and Garkovitch 1994: 21). In similar fashion, Hoffmann's (2004) cross-national study of social and economic factors that favour an increase in the number of endangered species represents the type of research that constructionists would not readily embrace.

Greider and Garkovitch conceptualise the idea of global environmental change as a type of 'landscape' and insist that by looking at how this landscape is symbolically created and contested, researchers are both 'incorporating and engaging' (Shove 1994). By doing so, they are contributing to the furtherance of a well-established school of thought in sociology and helping to forge a role for the discipline in the debates over environmental issues.

Second, social constructionism makes a valuable contribution to environmental policy-making by asking important questions about who makes claims for the existence of environmental problems and who opposes them, thus allowing us to situate environmental issues within relevant social and political contexts (Sutton 2004: 57). This is a task that has been more or less neglected or underplayed by other theoretical perspectives. In this regard, social constructionism makes a notable contribution in two ways (Davidson and Frickel 2004: 477–9): (1) by highlighting the ability of a particular discourse (for example, sustainable development) to become hegemonic and, hence, stifle debate; (2) by demonstrating how industry and state actors develop 'rhetorical strategies', especially during controversies, to convince the public that environmental problems are being competently addressed when in fact the opposite is true.

Transcending the nature/culture divide: co-constructionism and the analysis of socionature

As environmental sociology enters its fourth decade, one theoretical frontier can be located in various efforts to integrate constructionism and realism. The goal here is described as moving beyond or transcending the nature/culture dualism or divide and linking nature and society more closely within environmental sociology (Murdoch 2001).

One articulate advocate of this is Alan Irwin, who describes the social and the natural as 'actively-generated co-constructions'. Irwin sees sociology as poised to enter:

> a more exciting – and risky territory where existing categorizations – the social, the natural, the scientific, the technological, the human, the non-human – are seen to be fluid and contextually constituted rather than pre-determined.
>
> (2001: 178)

By deploying a co-constructionist strategy, Irwin argues, it is possible to avoid some of the perils of 'social reductionism' that haunt social constructionist analysis. This also allows us to ask some useful questions about contemporary environmental issues and problems. Is GM (genetically modified) food a social or environmental problem? Is the destruction of the rainforest a social or a natural disaster? Should the 'mad cows' at the centre of the BSE debate be construed as social or natural?

Such thorny questions inevitably open up a philosophical disagreement about the boundaries between the natural and the social. Eckersley (2004: 123–4) usefully puts this debate in perspective. In her view, it is a mug's game to declare that there can ever be one 'absolute' understanding of reality. Rather, any 'objective' knowledge that we have about our world is necessarily contingent, that is, it will invariably be 'historically and culturally specific, provisional and potentially always vulnerable to challenge and change'. At the same time, Eckersley cautions that this does not mean that there is no nature beyond the nature that we have socially constructed. On this point, she takes issue with Steven Vogel (1995) who denies 'extradiscursive nature' (that existing beyond human discourse) any independent existence. To take such a position, Eckersley charges, is to embrace a 'hyperconstructionism' that wrongly pushes social constructionism 'over the edge'. Her favoured approach, which she calls 'critical political ecology', avoids this extreme; it 'simply acknowledges that we don't have any *shared* access to reality other than through discourse' (p. 123).

In attempting to span the nature/culture divide, one prominent disagreement has revolved around the question of whether or not to grant non-human species any significant degree of agency (power to act). Vogel (1995) denies this on the basis that rocks or trees or butterflies cannot act as 'communicative partners' with humans. By contrast, the prominent French sociologist of science Bruno Latour (1999; 2000) refers to non-humans as *actants* who possess the power independently to do things that have either beneficial or dangerous consequences for humans, thus making them 'more than just a set of inert constraints' (Murphy 2004: 4). In this view, agency and power are conceptualised not as exclusive properties of individuals (humans or otherwise) but as the outcomes of networks composed of 'hybrids' of people, nature and technologies (Lockie 2004: 35).

In one well-known empirical illustration of this *actor–network theory* (ANT) approach, Callon (1986) cites the reluctance of scallop larvae in St. Brieuc, France, to anchor to artificial collectors immersed in the sea as evidence that the molluscs were 'contesting' the imposition of an emergent socio-technical network. The dilemma faced by Callon, Latour and other ANT thinkers is how to confer 'actor' status on scallops and other non-human agents without at the same time anthropomorphising (attributing human qualities to) them. Indeed, Callon (1986: 228, footnote 24) recognises this trap, advising his readers that 'the only thing that counts is the definition of their [the scallops'] conduct by the various actors involved'. This, however, ends up sounding suspiciously close to mainstream social constructionism, something from which actor–network theory has deliberately attempted to distance itself.

Despite its rising appeal as an exemplar for a new, refashioned 'ecological sociology', co-constructionism (and ANT) has not been universally celebrated. For example, David Bloor (1999), a noted critic of Latour, complains that ANT is all sizzle and no meat. That is, stripped of its distinctive, alternative vocabulary (emergent networks, actants, translation, enrolment), what remains does little to advance our empirical understanding of the socio-natural world. In fact, by abandoning the standard sociological lexicon, we risk losing the ability to explain the social beliefs and practices that compose science. While uncomfortable with Bloor's implied conclusion – that a clear boundary between 'nature' and 'society' be maintained so that sociologists can continue to investigate the social dimensions of environmental change – Murdoch (2001) nevertheless recognises that Latour and other actor–

network theorists have struggled to establish a co-constructionist mode of analysis. It still remains unclear, he observes, 'whether the ecological gains achieved by the theory outweigh the sociological losses' (p. 128). Finally, Yearley (2002a), in his review of Latour's book *Pandora's Hope* (1999) expresses considerable doubt about the efficacy of Latour's 'third way' for science. In that book, Latour devotes a chapter to his study of field scientists trying to work out whether parts of the Brazilian forest are advancing into the savanna or vice versa. Yearley concludes that Latour's attempt to sidestep the debate over whether parts of the Brazilian forest are *actually* advancing into the savanna or vice versa offers scant rewards or incentives for following his 'third way' (p. 167).

One worthwhile attempt to close the 'great divide' between society and nature (Goldman and Schurman 2000) that has recently attracted considerable attention is what has been called the analysis of *socionature*. First introduced by Callon (1986) in the 1980s, it has been developed conceptually and utilised pragmatically in a handful of empirical studies by the social geographers Eric Swyngedouw (1999) and Sally Eden *et al.* (2000).

Swyngedouw (1999: 446) describes socionature as 'a historical–geographical process' in which society and nature are inseparable, socially produced and transformable. In a figure entitled 'the production of socionature', he presents a dialectical model in which each of the component parts (language, discursive constructions, ideological practices, social relations, cultural practices, material practices, bio-chemical physical practices) are constantly swirling in and out from the production process itself. At the centre are 'hybrids' – part social, part natural. Swyngedouw applies this model to the case of modernisation and water landscapes ('waterscapes') in Spain from 1890 to 1930. He argues that the only way to understand this adequately is to explore how water, culture and social construction continually interweave. From this vantage point, modernity becomes a 'deeply geographical' project. It is also a contested one, in which the modernisers used their vision of the Spanish water map to inscribe a new set of power relations.

More recently, Swyngedouw (2004) has applied this model of socionature to the history of water politics in Guayaquil, Ecuador. The problem with both of these case studies is that it is difficult to see clearly what useful contribution the theoretical makes to the empirical. As Noel Castree (who is generally sympathetic to this type of analysis) writes in his review of *Social Power and the Urbanization of Water*:

> I find the book's dialectic approach to 'socionature' a tad too generous theoretically ...the rich empirical detail of the book is not always well connected to the theoretical statements of the early chapters. The salutary focus on institutions, class fractions and the like in Part 2 often seems a little detached from the grand abstractions of Part 1.
> (2005: 1471)

In other words, it is not very clear what the socionature model can explain here that a political economy or 'political ecology model' (see Chapter 4) cannot. As Sutton (2004: 74) observes of co-constructionist models in general, 'these alternatives remain closer to the constructionist pole than the realist one and do not really build on the effectivity of the natural world on social life'.

3 Environmental discourse

In recent years, discourse analysis has emerged as an increasingly influential method for analysing the production, reception and strategic deployment of environmental texts, images and ideas. Although closely identified with social constructionism, nonetheless, discourse analysis has been practised with good results by subscribers to other 'schools' of environmental theory and research, most notably, critical theorists, political ecologists and international policy analysts.

Hajer (1995: 264) defines discourse as 'a specific ensemble of ideas, concepts and categorizations that is produced, reproduced and transformed in a particular set of practices and through which meaning is given to physical and social realities'. Or, put more succinctly, discourse is an interrelated set of 'story-lines' which interprets the world around us and which becomes deeply embedded in societal institutions, agendas and knowledge claims. These story-lines have a triple mission: to create meaning and validate action, to mobilise action, and to define alternatives (Gelcich *et al.* 2005: 379).

Discourse is the most general category of linguistic production and subsumes a number of other tactics and devices including narrative (the writing and telling of stories) and rhetoric (see Chapter 5). Some rhetoricians have drawn the ire of critical realists by insisting that we can *only* conceive of nature and the environment through the discursive language that we have developed to talk about the natural world. However, a more temperate view is that the environment as it exists in the public policy sphere is the product of discourse about nature established by scientific disciplines such as biology and ecology, government agencies, bestselling books such as Rachel Carson's *Silent Spring*, and the messages disseminated by environment activists (Herndl and Brown 1996: 3).

Discourse analysts have also been criticised for overstating the importance of discourse in environmental politics and policy-making. Hajer (1995: 6), for example, insists that interests are constituted *primarily* through discourse, thereby excluding other institutional practices and institutions. The politics of discourse, he maintains, is not merely about 'expressing power-resources in language but it is about the actual creation of structures and fields of action by means of story-lines, positioning, and the selective employment of comprehensive discursive systems' (p. 275). Lidskog (2001) takes him to task for this, arguing that discourses are by no means the only determinant of social life. The discursive dimension, he points out, is 'only one of many that are relevant to sociological analysis' and, therefore, it is problematic to claim discourse analysis, as useful as it can be, effectively constitutes a 'general approach to environmental sociology' (p. 124).

Studying environmental discourse

Within environmental studies, discourse has been visualised in a variety of ways, ranging from a 'story-line' that provides a signpost for action within institutional practices (Hajer 1995) to a social movement 'frame' that enables the practices of environmental movement organisations (Brulle 2000), to an environmental 'rhetoric' constructed around words, images, concepts and practices (Myerson and Rydin 1996).

One basic attempt to organise the analysis of environmental discourse comes from Herndl and Brown (1996). Their 'rhetorical model for environmental discourse' takes the shape of three circles, each of which is located at the tips of a triangle. At the top of the triangle is what they call *regulatory discourse* – disseminated by powerful institutions that make decisions and set environmental policy. Nature here is treated as a resource. At bottom right of the triangle is the *scientific discourse* where nature is regarded as an object of knowledge constructed via the scientific method. Policy-makers routinely ground their decisions here, relying in particular on technical data and expert testimony. Finally, directly opposite this on the bottom left is *poetic discourse* that is based on narratives of nature that emphasise its beauty, spirituality and emotional power. Nature writing is one example of this. Herndl and Brown stress that these three powerful environmental discourses are not mutually exclusive or pure, however, and often end up being mixed together. In such cases, what we best look for are 'dominant tendencies' (p. 12).

Another effort directed at the classification of environmental discourses is Brulle's (2000) typology of discursive frames adopted by the US environmental movement. Drawing on the environmental philosophy literature and on his detailed reading of the history of American environmentalism, Brulle came up with nine distinct discourses: manifest destiny (exploitation and development of natural resources gives the environment value that it otherwise lacks); wildlife management (the scientific management of ecosystems can ensure stable populations of wildlife remain available for leisure pursuits such as sport hunting); conservation (natural resources should be technically managed from a utilitarian perspective); preservation (wilderness and wildlife must be protected from human incursion because they have inherent spiritual and aesthetic value); reform environmentalism (ecosystems must be protected for human health reasons); deep ecology (the diversity of life on earth must be maintained because it has intrinsic value); environmental justice (ecological problems reflect and are the product of fundamental social inequalities); ecofeminism (ecosystem abuse mirrors male domination and insensitivity to nature's rhythms); and ecotheology (humans have an obligation to preserve and protect nature since it is divinely created). Brulle argues that this multiplicity of discourses has resulted in the fragmentation of the US environmental movement, preventing it from speaking with a single, unified voice to a wise national audience. Adherents of each discursive frame talk past each other 'in a process of mutual incomprehension and suspicion' (p. 273). As do Schnaiberg and his entourage (see Chapter 2), Brulle concludes that there can be no meaningful environmental action without real structural change. This is unlikely to occur as long as discourses about the environment continue to block or mask the social origins of ecological degradation and proclaim a coherent vision of the common environmental good.

A third work that explicitly utilises the typological method is John Dryzek's (2005) book *The Politics of the Earth: Environmental Discourses*. Here, Dryzek identifies four main discourses: survivalism, environmental problem solving, sustainability and green radicalism. He organises these along two dimensions: prosaic vs. imaginative and reformist vs. radical. Prosaic dimensions are those that require action but do not point to a new kind of society, while imaginative departures from the long-dominant discourse of industrialism seek to dissolve old dilemmas and refine the relationship between the economic and the environmental. Each of these can be either reformist (adjusting the status quo) or radical (requiring wholesale transformation of the political–economic structure). According to this typology, problem solving is prosaic/reformist; survivalism is prosaic/radical; sustainability is imaginative/reformist; and green radicalism is imaginative/radical. Each of these four types is, in turn, subdivided. Problem solving, for example, comes in three forms: administrative rationalism, democratic pragmatism and economic rationalism, while sustainability has two flavours: sustainable development and ecological modernisation. For the most part, this typological exercise is helpful, although at an empirical level it requires some discriminating judgement calls on the part of the analyst as to what is imaginative and radical and what is not.

There are many other books and articles, of course, that discuss environmental discourses but do not propose typologies. Two of the best known of these deal with specific 'policy' discourses: Maarten Hajer's (1995) detailed analysis of the social construction of an ecological modernisation discourse on acid rain in Britain and the Netherlands in the 1980s and 1990s and Karen Litfin's (1994) account of changing international discourse about global ozone layer depletion in the 1980s. Killingworth and Palmer's (1996) article on 'apocalyptic' environmental discourse spans the period from the publication of Rachel Carson's *Silent Spring* in the 1960s up to more recent debates over global warming and climate change.

More recently, Craig Calhoun (2004) has identified a discourse of 'complex emergencies'. A discourse of emergencies, Calhoun tells us, is central to international affairs and now is the primary term for referring to a range of catastrophes, conflicts, and settings for human suffering. As such, it serves to organise a cluster of gradually developing, predictable and

Table 1 Typology of key environmental discourses in the twentieth century

	Discourse		
	Arcadian	*Ecosystem*	*Justice*
Rationale for defence of environment	Nature has priceless aesthetic and spiritual value	Human interference in biotic communities upsets the balance of nature	All citizens have a basic right to live and work in a healthy environment
Iconic books	*My First Summer in the Sierra*	*Silent Spring* *A Sand County Almanac*	*Dumping in Dixie*
Primary nesting place	Back to nature movement	Biological science	Black churches
Key alliance/fusion	Preservationists and conservationists	Ecology and ethics	Civil rights and grassroots environmentalism

enduring events and interactions into a 'crisis' that is 'sudden, unpredictable and short-term'. This constitutes, Calhoun says, 'a discursive formation that shapes both our awareness of the world and decisions about possible interventions into social problems' (p. 376).

Building on this prior work, in this chapter I offer up my own typology (see Table 1). As is the case with Brulle's nine discursive frames, the three environmental discourses presented here (Arcadian, Ecological, Environmental Justice) follow a rough chronological order, as each first rose to prominence at a different stage in the history of the environmental movement. In common with Herndl and Brown's model, a distinguishing characteristic is the predominant 'motive' or 'justification' for the environmental action.[1]

I begin with an account of the emergence in the early twentieth century of *Arcadian discourse*, which, in Herndl and Brown's terms, would be described as 'poetic discourse'. In contrast to the other three, Arcadian discourse peaked before the advent of the modern environmental movement in the early 1970s. Even so, the nature protection movement of the late nineteenth and early twentieth centuries acted as 'the advance guard of environmentalism' (Killingsworth and Palmer 1996: 43, note 4) and thus significantly shaped contemporary perceptions and views.

A typology of environmentalist discourse

Arcadian discourse

Writing in the Common Ground column of the British newspaper the *Guardian*, Robert Macfarlane (2005) recently offered a thoughtful elegy for the wilder landscapes of the British Isles. Every day, he observes, millions of people find themselves deepened and dignified by their encounters with these landscapes. Macfarlane knows this because he has come upon testimonies in the form of grafitti, memorabilia and even poems tacked up on walls. While distancing himself from those who regard wild landscapes as 'a site for the exercise of middle-class nature sentiment', nevertheless he urges his readers to rediscover the tradition of nature writing that slipped from view a half-century ago. This is vital because such landscapes are rapidly disappearing in what novelist John Fowles has called the era of 'the plastic garden, the steel city, the chemical countryside'. In lamenting the near abolition of remoteness and celebrating its pleasures, Macfarlane is evoking what has come to be called an 'Arcadian discourse'.

Van Koppen (1998: 74–5) assigns three defining features to Arcadian discourse: externality, iconisation and complementarity. *Externality* means that Arcadian nature is constructed as something external to human society, or at least removed from everyday life in the city. *Iconisation* suggests that the image of nature in the Arcadian tradition is modelled on stereotyped visual images that become embedded in cultural memory. In earlier centuries these were to be found in Dutch and English landscape painting, but today they are associated more with photos of primordial wilderness settings such as the Amazon rainforest. Finally, the Arcadian tradition is best understood within the context of its *complementarity*. That is, it stands in counterpoint to the urban industrial society and to the social and all of the environmental ills attached to it.

In his instant classic, *Landscape and Memory*, Simon Schama (1996) observes that there have always been two kinds of Arcadia: one infused by lightness and bucolic leisure, the other darker and a place of 'primitive panic' (p. 517). While it is tempting to see these two landscapes of the urban imagination aligned against one another, Schama maintains that over the course of human history they have, in fact, been 'mutually sustaining' (p. 525). Much the same point has been made by the environmental historian William Cronon (1996) who has described the pivotal concept of the 'wilderness' as having its origins in two broad sources: the 'doctrine of the sublime' as conveyed in the work of nineteenth-century Romantic artists and writers such as Wordsworth, Emerson and Thoreau; and the more recent notion of the 'frontier' as proclaimed by the American historian Frederick Jackson Turner. The convergence of these two discursive elements accelerated and coalesced in the 'Back to Nature' movement in the late nineteenth and early twentieth centuries, thereby 'clothing the wilderness in a coat of moral values and cultural symbols that has lasted right up to the present day' (Hannigan 2002: 315).

Wilderness as a discursive invention: the 'Back to Nature' movement in early twentieth-century America

As Europe and America became increasingly urbanised at the close of the nineteenth century, views towards nature began to undergo a major transformation. In particular, the concept of 'wild nature' as a threat to human settlement which had long predominated gave way to a new, intensely romantic depiction in which the wilderness experience was celebrated.

The traditional image of nature and its inhabitants as frightening is reflected in much of our past and present 'mythical' literature. For example, wolves play a central role in fairy-tales such as *Little Red Riding Hood* and *Peter and the Wolf* and more recently, in the Disney film version of *Beauty and the Beast*, making the woods a dangerous place for children to wander alone. Similarly, readers are advised to keep out of the forest at night to avoid spectres such as the Headless Horseman in *The Legend of Sleepy Hollow*. Civilisation is depicted here as the conversion of untamed natural landscapes into a more refined pastoral setting. Note, for example, Tolkien's contrast in *Lord of the Rings* between the gentle, civilised, rolling vistas of the hobbit settlements and the wilder, darker world of the forest and mountains inhabited by walking trees, orcs and other threatening creatures.

This unfavourable attitude towards untamed nature was especially heightened during the settlement of the American frontier:

> Wild country was the enemy. The pioneer saw as his mission the destruction of the wilderness. Protecting it for its scenic and recreational value was the last thing frontiersmen desired. The problem was too much raw nature rather than too little. Wild land had to be battled as a physical obstacle to confront and even to survive. The country had to be 'cleared' of trees. Indians had to be 'removed'; wild animals had to be exterminated. Natural pride arose from transforming wilderness into civilization, not preserving it for public enjoyment.
>
> (Nash 1977: 15–16)

By the last part of the nineteenth century, however, a revised view of unmodified nature had emerged. Rather than a threat, wilderness was now seen as a precious resource. This view was especially strong in the United States where the frontier was on the verge of closing. In the Eastern portions of the country, natural landscapes were rapidly disappearing as urban growth proceeded. Urban expansion, in turn, seemed to produce a surfeit of noise, pollution, overcrowding and social problems. In this context, unspoiled natural settings took on a special meaning; that is, the stress of city living created a rising tide of nostalgia among the urban middle classes for the joys of country life and outdoor living.

Schmitt (1990) has identified a 'back to nature' movement that flourished in the United States from the turn of the century to shortly after the First World War. This movement or 'wilderness cult' (Nash 1967) encompassed a wide range of activities including summer camps, wilderness novels, country clubs, wildlife photography, dude ranches, landscaped public parks and the Boy Scouts. While it was not the only factor, this nature-loving sentiment played a significant role in the creation of the natural parks system. In the process, wild nature was transformed from a nuisance to a sacred value. As the Ecological Society of America's Committee on the Preservation of Natural Conditions wrote in the *Naturalist's Guide to the Americas* (Shelford 1926), the wilderness, like the forests, was once a great hindrance to our civilisation; now, it must be maintained at great expense because society cannot do without it (Schmitt 1990: 174).

It is quite clear from Schmitt's and other accounts that this back to nature movement and the 'Arcadian myth' that it promulgated was socially constructed. While its supporters had mixed motives, they generally shared a belief that a return to nature represented a more wholesome set of values from those to be found in the increasingly corrupt environment of the city. Claims about the virtue of nature were made in each of the major institutions of the day. Leading American educators such as G. Stanley Hall, Francis Parker and Clifton Hodge actively encouraged nature study in the schools as a means of counteracting urban vices and building character. Religious educators, convinced that Americans could best find Christian values out of doors, promoted a form of pastoral Christianity in a number of ways: nature sermons, outdoor church camps, sponsorship of Scout troops, and so on. Nature journalists published a steady stream of nature lore, essays, outdoor pictures and literary tales (e.g. Jack London's *Call of the Wild*) celebrating the lure of wild nature. The case for wilderness preservation was taken on by a clutch of new conservation organisations from the Sierra Club (founded 1892) to the Wilderness Society (founded 1935). This preservationist sentiment was especially strong among bird-watchers and ornithologists who participated in a series of crusades for over fifty years in both Britain and the US to protect wild birds from hunters, poachers, feather merchants and other enemies (see Doughty 1975).

The back to nature movement gained a number of prominent political and institutional sponsors. None was more important than Teddy Roosevelt who, as Governor of New York and then as President, became a staunch advocate of wildlife preservation. Another key supporter was David Starr Jordan, the first president of Stanford University, whose voice in support of nature study gave the movement credibility and prestige (Lutts 1990: 28). A number of important figures in the movement were based in public institutions: the American Museum of Natural History, the Smithsonian, the Carnegie Institution and the New York Zoological Society to which they were able to bring considerable resources – money, publicity, prestige – to their preservationist and other activities on behalf of nature.

It was from these institutions also that many of the key popularisers of nature protection originated. For example, the movement to save the redwoods contained several leading scientific popularisers of the day: Madison Grant,[2] a New York lawyer and author; Edward

E. Ayer, head of the Chicago Museum of Indian History; Gilbert Grosvenor, founder of the National Geographic Society; and Fairfield Osborn, a key figure in the growth of the New York Museum of Natural History (Schrepfer 1983: 41). Perhaps the highest profile populariser (next to Teddy Roosevelt) was William T. Hornaday, for many years director of the New York Zoological Society (Bronx Zoo) who was a major force in lobbying Congress to tighten hunting regulations. Hornaday, a tireless self-promoter, wrote several widely distributed volumes on wildlife preservation as well as numerous articles in the *New York Times* and other popular publications. John Muir, the founder of the Sierra Club, was a charismatic promoter of wilderness protection who waged the country's first nationwide environmental publicity campaign during the Hetch Hetchy controversy.[3]

Popularisers such as Hornaday and Muir, as well as other claims-makers within the broad back to nature movement, were highly successful in garnering media attention. In this age of magazines, nature study essays and outdoor adventures were frequently featured in *Outlook*, *The Atlantic Monthly*, *Forest and Stream*, *Saturday Review*, *National Geographic* and other popular periodicals. In addition, various campaigns initiated their own publications, some of which developed a large readership. For example, the bird preservation movement spawned *Bird Lore*, the *Audubon Magazine* and other similar periodicals. *Boy's Life*, a monthly picture magazine that capitalised on the growing popularity of scouting, sold a cumulative total of forty-one million issues from 1916 to 1937 (Schmitt 1990: 111). One environmental campaign, the crusade to save Niagara Falls (1906 to 1910) was waged primarily in the pages of American popular magazines, notably the *Ladies' Home Journal*; it resulted in over 6,500 letters written in support of the preservation of the Falls (Cylke 1993: 22).

The back to nature movement drew upon a deep wellspring of existing cultural sentiments and in turn created a number of readily identifiable symbols and icons: the horse, Black Beauty,[4] the California redwood trees, the Grand Canyon, Old Faithful geyser in Yellowstone National Park and even Smokey the Bear. Some of these were real, others fictional creations. Nonetheless, as Schmitt (1990: 175) notes, 'those who dealt in symbols and myths found the wilderness a major force in shaping American character'.

Ecosystem discourse

A second major discourse that has powerfully shaped how we regard nature and the environment is that centring on the notions of 'ecology' and the 'ecosystem'. Referring to Herndl and Brown's (1996) terminology, we could say that the dominant tendency here is 'scientific discourse', although, as we will see, in the 1970s this fused with a normative strain within the emerging environmental movement.

Ecology has a long history prior to its ascendancy as the cornerstone of the contemporary environmental movement. Worster (1977: xiv) observes that while the term ecology did not appear until the latter part of the nineteenth century, and it took almost another hundred years for it to become a household word, the idea of ecology is much older than the name. Nonetheless, the term was officially coined in 1866 under the name *Oecologie* by Ernst Haeckel, the leading German disciple of Darwin. By ecology Haeckel meant 'the science of relations between organisms and their environments'.

The full development of plant ecology owed more, however, to plant geographers, most notably the Danish scholar Eugenius Warming who published his classic work *Plantsomfund*

(The Oecology of Plants) in 1895. Warming's central thesis was that plants and animals in natural settings such as a heath or a hardwood forest form one linked and interwoven community in which change at one point will bring in its wake far-reaching changes at other points (Worster 1977: 199). This is, of course, a central message in the contemporary ecological outlook.

Bramwell (1989: 4) has hypothesised that two strands of ecology emerged from this period. One was an anti-mechanistic, holistic approach to biology that derived from Haeckel and the plant geographers; the other a new approach to energy economics that focused on scarce and non-renewable resources. Bramwell argues that when these two strands finally fused together in the 1970s, the modern age of ecology was born.

If Bramwell is correct, why did this fusion not take place earlier, notably as part of the back to nature movement at the beginning of the twentieth century?

One answer to this is provided by environmental historian Susan Schrepfer who demonstrates that throughout this period most natural scientists were blinded to the hardcore implications of ecological thinking because of a commitment to various theories of directed evolution, notably that of 'orthogenesis'. According to this paradigm, genetic change was neither random nor was it influenced to any great extent by the surrounding environment. Instead, it followed an orderly progress. It was not known what constituted the prime force behind this orthogenesis or 'straight-line evolution'; one popular explanation was that it was possibly hormonal, another that it was part of a 'cosmic design'.

Most of the leading scientific entrepreneurs of the back to nature movement – Henry Fairfield Osborn, John Merriam, Joseph Le Conte – believed in this directional evolution. As Schrepfer cautions, the scientists who led the wilderness movement from the 1890s through the 1930s rejected much of the content of social Darwinism in favour of a reform Darwinism which taught that human reasoning power liberated us from the survival of the fittest. Instead, humans were thought to have the power actively to engineer progress; for example, by fighting to save the wilderness. At the same time, humans were regarded as the highest product of directed evolution – an achievement made possible through technological innovation. It is not difficult to see how this assumption led to a fundamental optimism regarding science and technology and a reluctance seriously to question the orderly march of industrial progress.

Accordingly, it was unlikely that ecology would have any strong appeal to the preservationists who were at the scientific centre of the back to nature movement. Not only did they have an unwavering faith that technology would overcome any problem of finite resources but they regarded humans' ability to cope as irrevocably cast within the evolutionary design of nature itself.

Nevertheless, by the 1920s biological ecology was coming into its own. Two of the major figures in its development were Frederic Clements and Arthur Tansley who developed a distinctively twentieth-century branch of biology called 'dynamic plant ecology' or 'ecosystem ecology'.

Clements, a Nebraska scientist who spent most of his career as a research associate at the Carnegie Institution of Washington, is best known for his study of ecological succession. He visualised the process of succession as going from an embryonic ecological community to a more or less permanent 'climax community' that was in equilibrium with its physical environment. Once formed, it was difficult for potential plant invaders to compete successfully with established species within this climax community. However, a number of external environmental factors – forest fires, logging, erosion – might damage or destroy the climax and force succession to begin again (Hagen 1992: 27).

Tansley, a British plant ecologist, is generally credited with coining the term 'ecosystem' in the mid-1930s. He strongly opposed Clements' use of the word 'community' to describe the relationship of plants and animals within a certain locale, maintaining that it was misleading because it wrongly suggested the existence of a social order (Worster 1977: 301). Instead, he came up with the concept of the 'ecosystem', which he described in terms of an exchange of energy and nutrients within a natural system. Catton calls the ecosystem the most central and incisive concept in the foundation of modern ecology, especially in Tansley's original understanding of the term, which was meant to 'unify our perceptions of nature's units' (1994: 81).

Tansley was eclectic in his interests and friendships, having, among other things, helped the social philosopher Herbert Spencer revise his *Principles of Biology* and pursued an interest in psychoanalysis by studying briefly under Freud and writing a popular book on Freudian psychology (Hagen 1992: 80). He was also an entrepreneurial scientific leader who played an instrumental role in establishing the British Ecological Society in 1913 and served for twenty years as editor of the Society's *Journal of Ecology*.

McIntosh (1985) has depicted the views of Clements, Tansley and other scientific ecologists of this era as being somewhat ambivalent with regard to human society. On the one hand, there was an acknowledgement that ecology had much to contribute to the understanding of human affairs. Clements (1905: 16) observed that sociology is 'the ecology of a particular species of animal and has, in consequence, a similar close association with plant ecology'. Tansley (1939), in his second presidential address to the British Ecological Society, anticipated the establishment of a worldwide ecosystem 'deriving from interdependence' and stated that human communities 'can only be intelligently studied in their proper environmental setting'. While it is probably an exaggeration to state as did some that ecology was the scientific arm of the conservation movement (McIntosh 1985: 297–9), nevertheless many ecologists were individually active in conservation causes. Tansley himself contributed towards the campaign to establish nature reserves and later (1949) served as the first chair of the British Nature Conservancy. In the 1940s, he led efforts (mostly unfulfilled) among ecologists to establish research linkages with the four British forestry societies on the grounds that post-war plans for giant new forest plantations would cause soil fertility to suffer as well as introducing an alien feature into the aesthetics of the countryside (Bocking 1993: 92–3).

Yet at the same time, ecologists and their societies were somewhat nervous of becoming too involved in political or social issues, fearing that their scientific credibility would be damaged. Both the British and American ecological societies were reluctant to engage in overt advocacy of particular positions or in political lobbying (McIntosh 1985: 308). Any synthesis of animal and plant ecology with human ecology was discouraged by the failure of the Chicago School in the 1920s and 1930s adequately to conceptualise the field.[5]

By the early 1970s, ecology had become the theoretical cornerstone of the new and rapidly diffusing concern with the environment. Ecologists increasingly began to step outside their role as scientists to become major contributors to the environmental debate. A plethora of new terms were added to the English language; for example, ecopolitics, ecocatastrophe, ecoawareness (Worster 1977: 341). A British magazine *The Ecologist* became a centre of gravity for left-leaning environmentalists under the guidance of Edward Goldsmith. Ironically, all this occurred in the midst of a deep intellectual schism within the field of biological ecology between ecosystem ecology[6] and evolutionary ecology over the concept of 'group selection', a form of self-regulation that naturally checks population growth.

There are several key factors that explain the centrality of ecosystem ecology in the rise of environmentalism in the 1970s.

First, the language and logic of ecology was linked to rising concerns about radioactive fallout, pesticide poisoning, overpopulation, urban smog and the like to produce what appeared to be an inclusive scientific theory of environmental problems. Rubin (1994) has argued that the instrumental force in effecting this transformation was a small group of influential writers and thinkers – Rachel Carson, Barry Commoner, Paul Ehrlich, Garrett Hardin – who functioned as scientific popularisers. Carson, in her book *Silent Spring*, brought the concepts of ecology, food chains, the 'web of life' and the 'balance of nature' into the popular vocabulary for the first time. Using ecology as the explanatory linchpin, she simplified a variety of problematic relationships into one 'environmental crisis' (Rubin 1994: 45). Commoner (1971) systematised Carson's observations with his four laws of ecology: 'everything is connected to everything else'; 'everything must go somewhere'; 'nature knows best'; 'there is no such thing as a free lunch'. These laws may have oversimplified ecosystem ecology but they had enormous rhetorical power. Similarly, Garrett Hardin's (1978) metaphor of the 'tragedy of the commons' found a broad audience both within the academic world and outside.

Second, the fusion of ecology and ethics first achieved in Aldo Leopold's 'land ethic' was featured prominently. The land ethic was first proclaimed in his book *A Sand County Almanac*, published posthumously in 1949. It extended ethical rights to the natural world, which he regarded as a community rather than a commodity. In the 1950s, Leopold's work had a small but committed following in conservation circles but only became widely known after it was reprinted in 1968. Whereas the ecosystem had previously been largely a theoretical construct, albeit a dynamic one, now it was inculcated with moral significance. Human interference in biotic communities not only had particular effects, for example, forcing a new round of succession as Clements had suggested, but it was also defined as the wrong thing to do. This insight became especially significant with the rise of 'deep ecology' in the 1980s.

Finally, as Macdonald (1991: 89) has observed, by co-opting scientific ecology the environmental movement added considerably to its strength for two reasons. First, despite the fact that ecosystem ecology was considered to be a somewhat 'soft' discipline within the natural sciences, it nevertheless allowed environmentalists to claim the authority of science for their campaigns. Second, because of its holistic perspective, ecology attracted a variety of 'seekers' such as devotees of expanded consciousness, Zen Buddhism and organic food who might otherwise have had little interest in green causes. Combined with scientific ecologists, these newcomers created a potent political mix. In recent years, this alliance has been at best an uneasy one but in the early 1970s it brought the idea of an 'ecological threat' into the pervasive currents of alternative popular culture where journalists constantly trawl in their search for the emergence of new trends.

Ecology, then, was transformed from a scientific model for understanding plant and animal communities to a kind of 'organisational weapon'[7] which could be used to systematise, expand and morally reinvigorate the environmental. In the process, it acquired a new texture: more political, more universal and more 'subversive' (Sears 1964; Shepard 1969). While some scientific ecologists reacted negatively to this reconstitution of the concept, others supported it, arguing that the 'environmental crisis' demanded a new sense of social activism on the part of biological researchers. The latter became influential claims-makers, presenting a politicised vision in which the boundaries between nature and society were deliberately blurred.

Kinchy and Kleinman (2003) have discerned the existence of two competing discursive tendencies within contemporary scientific ecology – purity and utility. On the one hand, it is argued that ecology is a value-free, objective science with legitimate claim to expertise. At the same time, many academic ecologists have explicitly aimed to demonstrate the discipline's usefulness in the policy-making arena. The Ecological Society of America (ESA), the primary professional scientific society for ecologists in the United States, has attempted to deal with these pressures by undertaking various programmes and initiatives designed to reap the benefits of public engagement while asserting the value neutrality of the discipline. For example, in 1979, having concluded the credibility of ecology was being sullied by non-experts claiming to be ecologists, the ESA established a voluntary certification programme designed to enable ecologists to participate in public debates over environmental issues while protecting the autonomy and unique expertise of ecology as a whole (pp. 882–3).

Most recently, the meaning of ecology has once again undergone yet another reconstruction. Grassroots activists such as those found in the Chipko Movement in India and the Greenbelt Movement in Kenya have proposed a new alternative ecological perspective in which insight into ecosystem interrelationships is achieved by means of folk knowledge rather than scientific observation. Indigenous wisdom of this type is embedded in practices that preserve the environment in the long run. Alas, local ecological knowledge has been suppressed by the juggernaut of global economic development, which forces the poor off their ancestral lands and deprives them of the opportunities to follow sustainable practices (Clapp and Dauvergne 2005: 109).

This alternative knowledge system provides citizens' movements with 'the epistemological tools for the reconstruction of neopositivist science and for an alternative approach to the management of global ecological independence' (Breyman 1993: 137). In this context, ecology becomes part mythology, part popular science: a rallying point for opposition to the kind of environmental diplomacy that predominated at the Rio Conference. As such, it represents a fresh social interpretation of a 130-year-old concept even if it is one that might be unrecognisable to Haeckel, Warming and other pioneers of scientific ecology.

Environmental justice discourse

In the 1980s, a new set of 'discursive formations' emerged in the United States that differed dramatically from prevailing ones in their interpretation of environmental problems and priorities. Environmental justice thought, Dorceta Taylor (2000: 508, 566) observes, has emerged as a major part of the environmental discourse; in the short time it has been around it has 'altered the nature of environmental discourse and poses a challenge to the hegemony of the NEP'.

Environmental justice lays out a set of claims concerning toxic contamination in terms of the 'civil rights' of those affected rather than in terms of the 'rights of nature' (Nash 1989). Capek (1993) identifies four major components of this environmental justice frame: the right to obtain information about one's situation; the right to a serious hearing when contamination claims are raised; the right to compensation from those who have polluted a particular neighbourhood; and the right of democratic participation in deciding the future of the contaminated community. Each of these components represents a specific claim that has been rhetorically formatted in the language of 'entitlement' (Ibarra and Kitsuse 1993).

Whereas the concept of ecology was utilised in the 1970s to join together rising concerns about toxic pollution with an ethical concern for nature, environmentalism in the 1980s and 1990s underwent another transformation in which the central discourse is 'environmental justice'. This shift occurred primarily at the grassroots level both domestically and in the Third World. While some key figures in this movement have wanted to throw off the environmental label entirely,[8] others have framed their claims to justice and equity within the context of an environmental movement. Environmental justice activists have not totally abandoned the legacy of the previous two decades: Commoner's industrial–ecological critique, for instance, has been one theoretical referent for this alternative explanation of the roots of the environmental crisis. At the same time, concerns about resource conservation, wilderness preservation and pollution abatement are de-emphasised in favour of issues such as the uneven distribution of resources and development and the safety of minority workers.

In the introduction to a special issue of the journal *Qualitative Sociology* on the topic of social equity and environmental activism, Alan Schnaiberg (1993: 203) rues the failure of environmental sociology to consider social inequality. As early as 1973, Schnaiberg claims, he was writing about the political necessity of incorporating elements of social justice into any proposal for environmental action but that this message fell on deaf ears. This may in part reflect shortcomings within the field as Schnaiberg suggests, but it is also a reflection of what was happening within the environmental movement itself.

In both the United States and Britain, the mainstream environmental movement was (and continues to be) dominated by a relatively narrow set of concerns; for example, rural planning and wildlife preservation. These are said to reflect the white, middle-class membership of the main environmental organisations.

In the United States a number of health-related environmental inequities were exposed during the 1960s and 1970s but they rarely made it into the larger movement agenda. Gottlieb (1993) highlights the differential treatment given to three issues during this period: pesticide poisoning; the toxicity of lead; and uranium hazards.

For migrant farmworkers in California the explosion of pesticide use through the 1950s and 1960s created a number of health-related problems. In its successful campaigns for farmworker rights during these years, the United Farm Workers (UFW) under the leadership of the charismatic Cesar Chavez aggressively attempted to pursue the pesticide issue, initiating court action to obtain information about the chemical ingredients and to ban specific pesticides including DDT, and pressing to have pesticide-related health and safety language incorporated into UFW–grower contracts.[9] Aside from some limited assistance from the Environmental Defense Fund, the mainstream environmental movement generally avoided the question of human exposure to pesticides, focusing primarily on the impact of pesticides on wildlife, as did Rachel Carson.

During the 1960s, childhood exposure to lead paint became a significant local issue in a number of inner city communities in the United States. By 1970, Gottlieb (1993: 247) notes, dozens of inner city-based community groups and coalitions were organising to address lead paint issues, primarily in East Coast cities such as Rochester, Washington, New York and Baltimore. Aided by New Left-inspired groups such as the Medical Committee for Human Rights, the lead paint movement achieved significant visibility both locally and nationally. At this point, however, the emphasis shifted to lead levels in the air, especially as a result of the emission of leaded petrol (gas). Mainstream environmental groups such as the Natural Resources Defense Council and the Environmental Defense Fund that had previously avoided involvement in this issue put a priority on reforming the Clean Air Act, eventually forcing a ban on the sale of leaded petrol. The lead paint issue did not re-emerge until the late 1980s and by then the primary claims-makers were from alternative environmental groups within the social justice movement.

Starting in the 1950s, uranium poisoning began to affect thousands of transient mine and mill workers, prospectors and residents of communities living downwind of the uranium mines. This 'radioactive colonisation' (Churchill and LaDuke 1985) was concentrated among native American workers in New Mexico and Arizona. For example, a 1979 spill of radioactive tailings into the Rio Puerto in Northern New Mexico contaminated significant stretches of Navajo Indian lands. As Gottlieb (1993: 251) observes, the Rio Puerto spill occurred just weeks after the Three Mile Island accident, yet it received limited attention from policy-makers and mainstream environmentalists. Indeed, during the 1950s and 1960s conservation groups ignored uranium issues altogether because they were perceived as occurring far from the scenic wilderness sites celebrated as part of an Arcadian discourse. During the following decade, environmental groups were primarily concerned with nuclear power as an alternative energy choice, although the anti-nuclear movement had begun to organise. Only in the 1990s did some groups accord the toxic threat to Indian lands a higher priority.

In each of these three cases, mainstream environmental groups focused on separate though often parallel concerns defining them in 'environmental' rather than 'social justice' terms (Gottlieb 1993: 253). In constructionist language, they established 'ownership' of the problems on behalf of a primarily upper-middle-class or élite Anglo constituency. On a more general level, they focused mainly on regulation or containment rather than seeking to subvert the social order in order to bring about a form of social reconstruction which would benefit 'have not' constituencies (Hofrichter 1993: 7). Kebede (2005: 89) has discussed this within a Gramscian perspective, contrasting the national environmental groups who are more interested in perfecting existing hegemony as against members of Grassroots Environmental Justice Organisations (GEJOs) who are more inclined to 'launch questions that go further into the innermost socioeconomic arrangements'.

In what proved to be somewhat of an anomaly, the Conservation Foundation, an organisation whose brief focused largely on research and education, convened a conference in Woodstock, Illinois, in November 1972 to explore the themes of race, social justice and environmental quality. At this gathering, urban planner Peter Marcuse, son of the famed social philosopher Herbert Marcuse presciently warned participants that divorcing equity and social justice concerns from the environmental agenda threatened to create a permanent rupture (Gottlieb 1993: 253). It would be another decade-and-a-half, however, until this rupture started to reach the public eye and the environmental justice discourse started to attract attention.

Kebele (2005) distinguishes between GEJOs whose membership is predominantly from people of colour and those whose membership is largely composed of blue-collar whites. The former draw their discursive tone largely from the civil rights movement of the 1960s while the latter draw on a set of values and assumptions embedded in the wider American political cultural system, notably the concept of 'total justice'.

Emergence and growth

In the United States, the environmental justice movement did not emerge until the early 1980s. As Bullard (1990: 35) notes, this newfound activism 'did not materialize out of thin air nor was it an overnight phenomenon'. Rather, it was the result of a growing hostility by urban blacks in the US to the siting of toxic landfills, garbage incinerators and the like in neighbourhoods or communities with predominantly minority populations. In the 1970s this was confined largely to the local context but within the decade it spread to a wider theatre as the struggle for environmental equity was presented as a fight against 'environmental racism'.[10]

There are several key milestones in the emergence and growth of the environmental justice movement during this period.

In 1987, the Commission for Racial Justice of the United Church of Christ issued an influential report entitled *Toxic Wastes and Race in the United States* which documented and quantified the prevalence of environmental racism. The UCC report firmly established the 'grounds' for this claim, by setting out the magnitude of the problem in numerical terms. Among its findings was the revelation that three out of five black Americans live in communities with uncontrolled toxic waste sites. Furthermore, blacks were heavily over-represented in those metropolitan areas with the greatest number of such sites: Memphis, Tennessee; St Louis, Missouri; Houston, Texas; Cleveland, Ohio; and Chicago, Illinois, with over a hundred each. Hispanics, Asian Americans and native peoples were similarly over-represented in high-risk communities. This confirmed a study conducted four years earlier by the US General Accounting Office that reported that three of the four largest commercial landfills in the South were located in communities of colour.[11]

Also crucial in establishing the dimensions of environmental racism was the research of sociologist Robert Bullard. In 1979, Bullard, a professor at the predominantly black Texas Southern University in Houston, was invited by his wife, a lawyer, to conduct a study on the spatial location of all of the municipal landfills in that city in order to provide data for a class action lawsuit that she was arguing. Bullard confirmed that toxic waste facilities, not only in Houston but elsewhere in America, are most likely to be found in black and Hispanic urban communities. In a series of journal articles beginning in 1983 (many co-authored with Beverly Wright) and in his book, *Dumping in Dixie*, Bullard documented these environmental disparities and the mobilisation of the environmental equity movement. Even more than was the case for the UCC report, Bullard's work established the 'warrants' for this problem, arguing that action was justified in order to reclaim for minorities 'the basic right of all Americans – the right to live and work in a healthy environment' (1990: 43). Bullard has since become a key leadership figure, as indicated in 1992 when he was chosen by the Clinton administration to participate in the Presidential transition process as a representative of the environmental justice movement (Miller 1993: 132).

In January 1990, Bunyan Bryant and Paul Mohai, professors in the School of Natural Resources at the University of Michigan, organised the University of Michigan Conference on Race and the Incidence of Environmental Hazards with papers from twelve scholar-activists. Among the follow-up strategy of the Michigan Conference was a series of meetings in Washington with key government officials including William Reilly, Administrator of the EPA, and Congressman John Lewis (Bryant and Mohai 1992).

Third, under the sponsorship[12] of the Commission for Racial Justice, the First National People of Color Environmental Leadership Summit was held in October 1991 in Washington, DC. At this gathering, three strands of environmental equity were identified (Lee 1992); *procedural equity* (governing rules, regulations and evaluation criteria to be applied uniformly), *geographic equity* (some neighbourhoods, communities and regions are disproportionately burdened by hazardous waste) and *social equity* (race, class and other cultural factors must be recognised in environmental decision-making). Delegates to the Summit ratified a document, *Principles of Environmental Justice*, which sets out the ideological framework of the emerging environmental justice movement. Taylor (2000: 537–42) organises these principles into six thematic components that deal with ecological principles; justice and environmental rights; autonomy/self determination; corporate–community relations; policy, politics and economic processes; and social movement building. While grounded in the ecocentric principles espoused by the pioneers of the environmental movement (John Muir, George Marsh, Aldo Leopold), the principles also argue that people have a right to clean air, land, water and food and the right to work in a clean and safe environment. The document affirms the rights of people of colour to determine their own political, economic and cultural futures. It strongly opposes military occupation and exploitation of their land and calls for their participation as equal partners in the policy arena. To ensure environmental justice, the *Principles* call for strong social movement building, both nationally and internationally.

Gottlieb (1993: 269) credits the summit with advancing the environmental justice movement past a 'critical threshold in definition' both by ratifying a common set of principles and by identifying a new kind of environmental politics of inclusion.

Organisationally, the movement has been held together in a number of decentralised, loosely linked networks of umbrella groups, newsletters and conferences (Higgins 1993: 292) rather than the top-down, professionalised configuration typical of mainstream environmentalism. This has its roots in the formation in the early 1980s of several based 'anti-toxics' groups – the Citizens' Clearinghouse for Hazardous Wastes and the National Toxics Campaign. In the mid-1990s, the emphasis shifted somewhat from national to regional grassroots networks, as epitomised by the Southwest Network for Environmental and Economic Justice.

While the social construction of discourse and strategic framing were crucial in communicating the environmental justice message to supporters, these were not by themselves enough, Taylor (2000: 563–4) notes, to mobilise a strong base of supporters. Rather, the

EJM adopted several key recruitment strategies. Rather than try to build movement networks from scratch, organisers tapped into lines of pre-existing social relationships and networks, drawing from networks of people with past histories of social and political connections. In particular, they targeted people with strong institutional ties to churches, labour unions, universities, community organisations, federal agencies, legal institutions, grant givers, and mainstream environmental organisations. Having observed the growth throughout the decade of the 1990s of Federal Government offices, programmes and initiatives devoted to pursuing environmental equity, the latter (mainstream environmental groups) finally started collaborating with communities of colour and EJOs, 'slowly diversifying their staff and memberships, covering environmental justice issues in their magazines, launching environmental justice programs, and undertaking a variety of environmental justice initiatives' (Taylor 2000: 559). Pellow gives a good example of this changing relationship. During the 1970s and 1980s, many environmentalists in the Chicago area endorsed the growing waste-to-energy incinerator industry as a way of converting trash into a useful form. By the 1990s, however, they reversed their support for incineration, in large part because they observed activists in communities of colour and working-class neighbourhoods engaged in struggles to resist industry's efforts to site these facilities near their homes.

One mainstream environmental group that has signed on to the environmental justice agenda in a significant way is the Sierra Club. In 1993, the Sierra Club adopted its first environmental justice policy, stating that 'to achieve our mission of environmental protection and a sustainable future for the planet, we must attain social justice and human rights at home and around the globe' ('Joining together for justice'). Since then, the Club has undertaken a number of initiatives: providing organisational assistance to over 250 low-income neighbourhoods and communities in the US; hiring full-time environmental justice organisers in Detroit, Washington, DC, the Southwest and central Appalachia; awarding grants to help local groups organise and lead 'toxic tours', create community gardens and undertake public education programmes about the environmental damage caused by factory farms in the South; helping block or shut down polluting mines, incinerators and sewage treatment plants; and collaborating with Amnesty International to defend activists under threat from the state for speaking out on environmental issues.

In the early 1990s, the environmental justice movement expanded its charter to incorporate the exploitation of Third World peoples. Much of the interaction between grassroots activists from the United States and their counterparts in the South has taken place in the context of the United Nations; for example, at the Rio Summit and its preparatory meetings. Environmental justice activists from the US also participated in a 1992 meeting hosted by the Third World Network in Malaysia that focused on toxic waste. These networking activities with Third World activists have moved environmental justice leaders back on the path to a renewed ecological awareness. Vernice Miller, a co-founder of the group West Harlem Environmental Action, described environmental justice as 'a global movement that seeks to preserve and protect global ecosystems' (1993: 134).

More recently, Anand (2004: 15) has drawn a parallel between the themes of the American environmental justice movement and international environmental politics between North and South. In particular, he identifies inequities in the international arena relating to both procedural and distributive justice that are similar to the national politics of environmental justice in the United States. Just as the environmental justice movement in the US represents a backlash against the failure of government to address gross inequities in exposure to toxic dumps and other health hazards, there has been 'tremendous opposition to many international global agreements and efforts because they do not adequately reflect the interests of countries of the South' (p. 15). The Biodiversity Convention is a leading example of this (see Chapter 9). Furthermore, the power imbalances inherent in the global economic system lead to situations where lower income residents in the nations of the South are differentially impacted environmentally. Not only is this a case of differential exposure to industrial effluents and other pollutants, but it has also meant inequities in access to basic natural resources such as fuel and drinking water (see Chapter 4).

As in Herndl and Brown's (1996) rhetorical model of environmental discourse, the three environmental discourses discussed above should not be treated as being either static or mutually exclusive. Rather, they engage one another in dialectical fashion. For example, in Southern nations, environmental activists have successfully merged elements of the ecosystem and social justice discourses. Creating 'new imaginaries' helps energise local struggles and draw in a more diverse set of allies by 'giving the demands of subordinate groups a new claim to universality' (Evans 1992: 8–9). Over time, Dryzek (2005: 20) tells us, environmental discourses 'develop, crystallize, bifurcate and dissolve'. And, sometimes, they return in a different wrapping.

Accordingly, Arcadian discourse, whose zenith in America passed nearly a century ago, has re-emerged in recent decades in the form of a romantic and spiritual celebration of the Amazon rainforest. Thus, Slater (2002: 101) describes current images and accounts of the Amazon such as those that are used to attract eco-tourists as having a 'dual nature'. That is, they intertwine 'virgin' and 'virago' – traditional narratives of a 'lush, dark, exciting jungle' that is harsh and untamed and contemporary narratives of a fragile rainforest composing an intricate web of flora and fauna. This latter image, in turn, connects with ecological discourse that portrays the rainforest as 'both a storehouse of valuable commodities and a key to global environmental health' (p. 138). A final discursive layer imbues this exotic, biotic paradise with an extra layer of moral urgency by drawing on a justice frame to publicise the plight of rubber tappers (notably Chico Mendes who was murdered in 1989), Indian tribes (Kayapó, Yanomani, Machiguenga) and other indigenous forest dwellers who face displacement and decimation at the hands of ranchers, miners and other agents of development.

4 Discourse, power relations and political ecology

It is difficult indeed to talk about discourse these days without bringing in a discussion of power. In large part, this is due to the influence of the ideas of the French social theorist Michel Foucault (1979; 1980). In various essays and books dating from the early 1960s to the mid-1980s, Foucault transformed our theoretical understanding of power as well as putting discourse analysis on the academic agenda. Most analytic perspectives in the humanities and social sciences that employ the concept of discourse 'have Foucauldian elements in terms of viewing discourses as something that defines what is meaningful and how it exercises power' (Gelcich *et al.* 2005: 379).

Foucault dismissed the previously paradigmatic notion that power necessarily resides permanently in institutions, notably the state. Rather, he conceptualised power as being embedded in social relationships. As such, it is not just a weapon wielded by the ruling class but a fundamental feature of everyday human interaction. As Hindness (2001: 100) has phrased it, power is manifested 'in the instruments, techniques and procedures that may be brought to bear on the actions of others'. This can range from the power of a president or prime minister to control the agenda of a national news conference, to the power of one partner in a marriage to control the choices of television programmes made by his or her spouse.

Power may be everywhere but relationships of power are rarely symmetrical and wholly democratic. Foucault makes an important distinction between power and domination. The latter refers to asymmetrical relationships of power in which the subordinated party has a negligible chance of exercising his or her will. Whereas power relationships are often unstable and reversible, domination means that these relationships are less fluid and less open to negotiation.

In the case of interpersonal power relations, one individual may hold power over another due to superior physical strength, attractiveness to others, rhetorical abilities, higher income or social status, or a more extensive network of political contacts (Scott 2001: 135). Where power is structured around formal institutions such as the state and the corporation, other means are required. In such cases, Foucault believed that power is exercised not so much through naked force and physical coercion as through the ability to shape the process of socialisation. This is much more effective because it reduces resistance while internalising consent. It is at this point that discourse becomes important.

Discourse provides institutions with a powerful means of incorporating individuals into relations of domination. Foucault regarded this as central to a process of social control that he labelled 'discipline'. Increasingly, he observed, this occurs under the supervision of 'experts' who are empowered through their stranglehold on scientific and technical forms of discourse (p. 92). While Foucault was primarily concerned with the exercise of discipline within total institutions such as prisons and psychiatric hospitals, his insights about the close relationship between discipline and expertise can just as easily be extended to the domains of science and environmental risk determination.

In the ongoing cultural contest in which discourse is shaped, some players possess more resources than others. In their cross-national study of talk about abortion, Ferree *et al.* (2002) coin the concept of a 'discursive opportunity structure', which they define as the 'complex playing field [that] provides advantages and disadvantages in an uneven way to the various contestants in framing contests' (p. 62). Here, they are drawing on the social movements literature, combining the framing approach of Snow and Benford with the political opportunity perspective of Charles Tilly and others. Ferree and her colleagues are especially interested in the power exercised by large institutions such as the mass media, the judiciary, the churches, the political parties and social movement organisations and how this shapes the process of producing (shaping) abortion discourse (Monteiro 2005: 160).

Control over discourse production also carries with it the power to 'delimit both the actors that can legitimately engage in politics and the issues that are subject to debate' (Davidson and Frickel 2004: 478). This is nicely illustrated by a case study by Carolan and Bell (2004) of a dioxin controversy that flared in 2000/2001 in the small Midwestern city of Ames, Iowa.

The conflict began with a report commissioned by the North American Commission for Environmental Cooperation by the internationally known scientist and environmentalist Barry Commoner and colleagues at the Center for the Biology of Natural Systems in New York. In his report, Commoner attributed accumulations of dioxin, a toxic chemical, in the Inuit community of Nunavut in the Canadian Arctic to 'drift' from a handful of major polluters in the United States. One of those cited was the waste incinerator attached to the municipal power plant in Ames. AQLN, a local activist group that had previously had some success in turning a local quarry into a park and water supply reservoir, organised a campaign of opposition. Commoner himself was brought to Iowa State University to deliver a lecture on 'Globalization and the Environment'. In response, the City commissioned Robert Brown, an Iowa State engineering professor to conduct a study as to whether the power plant should be tested. Citing scientific inadequacies in the Commoner report (lack of data directly connecting incinerator emissions with Arctic dioxin build-up; apparently incorrect information on plant use and construction), Brown recommended against in-plant testing.

Citing Foucault's observation that social relations are also relations of power, Carolan and Bell (2004: 287) stress that the city government and the university engineers effectively controlled the public debate. AQLN's 'threat to public health' frame never effectively competed with an official 'the Ames power plant is safe and a source of community pride' frame. Other than Commoner's, the activists' voices were rarely heard in the local press. As relative newcomers to the city, AQLN members were not well integrated into local community networks. By contrast, 'those with access to the dominant social networks also have an avenue through which to express their opinions and have these options heard, all of which has bearing on how others perceive them as speaking the truth' (ibid.).

Discourse and political ecology

Discourse and discursive clashes play a central role in recent scholarship that follows the terrain of what has been called the *new political ecology*. This takes the form of 'locality-based studies of people interacting with their environments' (Goldman and Schurman 2000: 568). Contemporary political ecology, Evans (2002: 8) explains, 'arose out of a dissatisfaction with traditional versions of ecological arguments, which tended to ignore the dilemmas of people whose livelihood depended on the continued exploitation of natural resources'.

Researchers following a reformulated political ecology approach have focused in particular on environmental struggles related to North–South relations. In this context, the term 'South' is used not only in a geographic sense to refer to Asia, Africa and Latin America, but also to reflect 'the common experiences of people in these countries as a result of historically determined social and economic conditions resulting from their colonial and imperial past' (Anand 2004: 1).

Escobar (1996) has argued persuasively that capitalist development today is routinely sheathed in seemingly beneficial discourses such as 'sustainable development' and 'biological conservation'. This is easy to do because their meaning is, at the very least, ambiguous. The underlying purpose here, however, is always to 'capitalise nature'.

Goldman and Schurman (2000: 570) identify two ways in which political ecology scholars have usefully employed discourse analysis: (1) as a method of understanding alternative discourses on nature, the environment and environmental degradation and how these clash with dominant discourses imposed by the state, Northern environmental movements, and transnational NGOs; (2) as a means of exploring and exposing the power relations embodied in national and global conservation agendas.

As we have seen, formulating and communicating ecological discourse is not restricted to those in power, although the latter have the upper hand in making their discourses dominant. There is a growing literature in the social sciences that discusses 'alternative' discourses on nature and the environment that flow from the grassroots. These discourses are rarely unopposed, however, since they inevitably challenge the state and other claimants to local land and natural resources.

Paul Ciccantell (1999) has made the important point that the discursive struggle in a socially remote extractive periphery such as the Brazilian Amazon is usually a matter of powerful external actors imposing their 'definitions' over the objections of indigenous groups. He illustrates this with a case study of the Tucuruí dam, built on the Tocantis River in the eastern Amazon in the 1980s as a joint venture between the Brazilian government and the Japan International Cooperation Agency, a Japanese government agency that was working with a consortium of over thirty Japanese aluminium producing, consuming and trading firms. The Tucuruí dam effectively cut off all river transport, forced the relocation of 20,000–30,000 people, and transformed local ecosystems thereby causing threats to human health (e.g. malaria), local climatic changes, the proliferation of new plant and animal species, and the decline of existing species (Ciccantell 1999: 306).

In this and other similar cases, three principal meanings formulated by the Brazilian military government in the 1960s and 1970s prevailed: the region's rivers were an obstacle to road-building and colonisation efforts; the rivers were a source of hydroelectric power for the raw materials processing industries and growing regional population centres; and waterways were access routes for oceangoing ships to export raw materials at low cost. Ciccantell stresses that a 'pluralist' model of social construction such as that which prevails in the United States and Europe, in which competing groups seek to define issues in terms that support their interests, does not normally operate here. Rather, the discursive process is dominated by regional and national economic and political élites that are able to impose their definitions, even in the face of organised public opposition (p. 296).

Another important depiction of these discursive struggles can be found in Nancy Peluso's reporting of her research in the 1990s in Java and in the western interior of Borneo both of which are part of Indonesia.

In an earlier study, Peluso (1992) focused on the struggle over the teak forests of Java. The Indonesian state tried to assert resource control over both the forest and its indigenous inhabitants by applying the modern legal constructs of 'property rights' and 'criminality' (those who violated the former). In response, the forest dwellers developed 'a counter-discourse on what is a fair, legal and legitimate use of the forest' (Goldman and Schurman 2000: 569).

In her later research, conducted in the province of West Kalimanata, Peluso (2003) concentrates on strategies used by local people to counter official government mapping exercises that are justified on the grounds that they are an ongoing part of 'territorial resource management'. As part of their efforts to mount 'counterclaims or reclaims to contested or appropriated resources' (p. 232), villagers engage in 'countermapping'. This is a technique by which traditional land and resource claims, many of which go back to a time before the Dutch colonised Indonesia, are expressed in a contemporary format. Using locally drawn 'sketch maps' that reflect local custom or practice, the 'mappers' – local activists sometimes assisted by international consultants – develop high-tech maps that are used to make claims to government and large international NGOs.

Peluso argues that this may be practically sound, but it contributes to the emergence of a new 'hybridized discourse' in which 'common rights in long-living trees, held communally by multiple generations of heirs, are slowly being replaced with a notion that property rights in land supersede or dominate all forms of property in trees and other territorial resources' (p. 233). In countermapping, NGOs and others are utilising 'state tools' and buying into state discourses of 'territorialisation'. They are also opening the door to several undesirable possibilities. Once rights to resources are mapped or documented, the state gains a certain power over these resources and the people claiming them, becoming both 'a recognized arbiter and mediator of both access and rights'. Furthermore, the conditions are established for increased community conflict, especially in regions that will likely see increased migration and intermarriage in the future.

A second way that discursive struggles may arise in the countries of the South is in response to attempts by national and international NGOs (INGOs) to impose their agendas and viewpoints on indigenous peoples. In particular, this is manifested in the tendency of INGOs selectively to take fragments of localised knowledge and translate these into the 'global' discourse of science (Dumoulin 2003: 593–4). This has been very much the case with the issue of 'biodiversity loss' (see Chapter 9) where, until relatively recently, environmentalists from abroad were committed to establishing biosphere reserves and other protected areas, usually at the expense of local people.

Using Mexico as an example, Dumoulin (2003) demonstrates how national and international ENGOS (environmental non-governmental organisations) have successfully framed the protection of 'indigenous' knowledge within the framework of a new, global-oriented approach to nature conservation. In particular, an 'epistemic community' (see Chapter 7) composed of ethnobiological experts with similar academic backgrounds and common values (enhancement of cultural and biological diversity for the future of humankind) have effectively exerted their influence in international arenas of power. They have done so by creating a cognitive framework, disseminating information and lobbying politically. The key group here is the International Society of Ethnobiology (ISE) in concert with the leaders of the Amazon Alliance, the Forest People Programme, the World Rainforest Movement and Cultural Survival. Under the direction of its founder, Darell Posey, the ISE has been particularly successful at securing positive media coverage and in ensuring that the issue of biodiversity-related indigenous knowledge is on the international environmental agenda (see Dewar 1995).

Environmental NGOs are not always, however, the most faithful translators of Western conceptions into Western discourses. As Roué (2003) points out, they are often no closer to the socio-cosmic view of indigenous peoples than were the resource development forces that preceded them. At best, they 'enable marginal populations, deprived of political power to acquire some at least, and to enter into communication with the centre and the dominant society' (p. 620). This is complicated further by the fact that ENGOs, by their very nature, are concerned not so much with the people to which they relate as 'the natural environment they inhabit, which is often perceived as wild and in need of protection' (p. 621). On occasion, this can lead to misunderstanding and conflict.

One particular point of difference has been over the image of local people as noble 'defenders of nature' employing their ancestral wisdom to protect non-renewable resources. In the case of indigenous people in both North and the South this is only true to a certain extent. Some traditional inhabitants, for example the Kayapó of the Amazon, are in fact quite entrepreneurial and are willing to sell their gold and timber for the right price (Dewar 1995; Slater 2002; Conklin and Graham 1995). As Slater (2002: 150) has noted, indigenous people in rainforest areas have proven most adept at borrowing vocabularies of human rights and environmental protection for their own ends. So too have tribal leaders in parts of Canada. In response, some disenchanted conservationists have concluded that sustainable development is impossible and that rainforests can only be protected through excluding all humans, including local people who dwell there. This has sparked a renewal of conservation programmes wherein tracts of land with relatively untouched natural ecosystems are purchased with public donations and fenced off as 'nature reserves' in order to keep them 'pure'.

The lesson here, I think, is not that rainforest dwellers or other local populations in the South are necessarily 'frauds'. Neither should it be concluded that the threats posed to the environment by mining, forestry, road-building, corporate agriculture and urban sprawl are to be discounted. Rather, it should tell us that discourses that frame the situation in simplistic terms as a conflict between 'conservation' and 'exploitation' with local inhabitants assuming the role of 'environmental defenders' are best treated with caution.

The case of water privatisation

These elements of discourse, power, conflict and ecology come together in a relevant fashion in what promises to be one of the leading environmental issues of the early twenty-first century – the privatisation of global water services.

In contrast to a good many of the socially constructed environmental risks that are discussed throughout this book, there is generally a consensus on the nature, source and scope of problems relating to the supply of drinking water worldwide. This has been described as a rising incidence of 'water stress'. Water stress is triggered when the rate of consumption outstrips available supply. During the twentieth century, water use grew at more than twice the rate of population increase. About a third of the world population lives in countries that are currently experiencing moderate to high water stress, and according to World Meteorological Association estimates, this is likely to rise to two-thirds by the year 2025. Polluted water is said to contribute each year to the death of about 15 million children under five years of age (Broad 2005: D1). Although there are regions that suffer from severe drought and overall supplies are declining, often the problem is less one of water availability than it is the inability to make it available to consumers.

According to Lipschutz (2004: 191), the water supply problem is twofold. First, in many parts of the world the infrastructure of water storage and delivery is inadequate, especially in cities that are experiencing rapid expansion. Second, agriculture, industry and cities are all competitors for the same limited water supplies. While this is most common in the nations of the South, it may also be observed in high-income regions such as the US Southwest, where the waters of the Colorado River are increasingly insufficient to supply a wide range of conflicting needs ranging from corporate agriculture in California, casino resorts in Nevada and residential lawn watering in Arizona.

In some cases, these problems have been exacerbated by corruption, poor management practices and conflicting government priorities. For example, Keck (2002) has presented a detailed analysis of failed efforts over much of the twentieth century in São Paulo, Brazil, to establish a dependable supply of clean drinking water and reliable sewage disposal. Partly, this was a matter of administrative gridlock where overlapping agencies refused to surrender their claims. Also, it was a sign of Brazil's pervasive 'clientalism' whereby megaprojects, often ill advised, were nevertheless approved in order to supply those in power with political exchange capital. Finally, it reflected a 'developmentalist' vision that privileged the quantity of water over the quality, since the latter contributed positively to hydroelectric power generation, a key to rapid economic growth.

Whatever the specific ensemble of causes, it is generally agreed that the increasing pressure on water resources and increasing demand have made an increasing level of investment in water supply mandatory. In the past decade (1995 to 2005), the World Bank estimated that a whopping $US 60 billion per year needed to be invested in the water-supply sector to ward off water stress (Haarmeyer and Mody 1997, cited in Bakker 2002: 769).

While there may be a broad, if not perfect, consensus on the diagnosis of the problem, this disappears when it comes to the question of how best to address it. Haughton (2002: 803) identifies three main positions here: pro-privatisation, improved public sector provision and extended community-based provision. Each of these brings with it a distinct discursive and ideological position.

The first of these has been formulated in the context of the rise of 'neoliberalism' in Britain and America in the 1980s and 1990s. Neoliberalist doctrine stipulates that the role of the state must shrink and change. This is commonly described as the 'hollowing out of the state'. Whereas formerly government was directly responsible for key social and physical services, now it is expected to become a 'strategic enabler' and marketer. For example, the proper strategy for large cities, caught in a seeming downward spiral of deindustrialisation and economic decline, is to embrace a new form of doing business that has become known as 'entrepreneurial governance'. Basically, this involves a shift in the role of local government from providing services such as garbage collection and welfare to becoming promoters, pitching the opportunities and attractions of the community to everyone from tourists to sports team owners (Hannigan 2005: 257). London's successful bid for the 2012 Summer Olympics represents the kind of entrepreneurial activity that neoliberal thinkers believe best showcases the new role of the state.

In the utilities sector, this neoliberalist agenda has manifested itself in the privatisation of gas, electricity, telecommunications, and, increasingly, water. To justify this, the state has been 'discursively constructed as a site of crisis' (Haughton 2002: 792). Publicly run utilities have been depicted as being overstaffed and grossly inefficient. The only answer, it is said, is to download these responsibilities to private firms. Government is not expected to completely withdraw, but rather to assume an emergent role as a watchdog or regulator, ensuring that private utility firms 'operate towards government priorities for economic efficiency, social equity, and environmental responsibility' (ibid.). In the UK, a massive water privatisation programme was introduced by the Conservative government at the end of the 1980s, wherein the ten regional water authorities in England and Wales were turned over to private companies.

By the late 1990s, the increased marketisation or privatisation of water management meant that a handful of giant conglomerates were well on the way to acquiring a significant share of the world drinking water systems. In 2004, multinational companies ran water systems for 7 per cent of the world's population, a figure that is projected to grow to 17 per cent by 2015. Private water management is estimated to be a $200 billion business, with the potential to be worth $1 trillion by 2021 (Luoma 2004: 53). This may even be higher, depending on the extent to which the water market in China opens up to private and foreign capital. Since the 1990s, more than fifty large international water companies have entered the Chinese market, with the largest presence being exerted by two French companies, Suez Lyonnaise des Eaux and Véolia Water, and a British company Thames Water (World Environmental Journalists 2005).

As Haughton has noted, multinational water companies have generally targeted cities in Latin America and Asia where the risk of market investment is perceived as being lower and the physical infrastructure is already in place. They are less inclined to look to the poorest countries and rural areas where a drinking water system would needs be constructed from scratch and the ability of consumers to pay full market costs without significant subsidies from the state is low. As a result, the private sector has been relatively slow moving into Africa. Several recent World Bank-funded projects in Ghana and Tanzania have collapsed amid acrimony. And in South Africa, water privatisation has become a political flashpoint and 'a source of conflict, division and distrust' (Carty 2003).

In the poorer countries of the South, one of the most important promoters of water privatisation has been the World Bank. As a powerful advocate of neoliberal 'reforms', the World Bank has the power to persuade countries to introduce private operators, insofar as it controls their borrowing capacities. This is justified on the grounds that the economies introduced by privatisation will free up state expenditures, with the savings being applied to address more urgent social priorites. The Bank, and its sidekick, the International Monetary

Fund, may also, of course, hope that pressuring these nations into privatising water delivery will lead to a more faithful repayment of development loans (Luoma 2004: 54).

While some NGOs, notably the Environmental Defense (formerly the Environmental Defense Fund) in the United States have supported privatisation on the grounds that free markets in water are a means of conserving a valuable resource (Lipschutz 2004: 192–3), most have opposed it. Some have argued in favour of Haughton's third option: local, community-based alternative management systems that favour private delivery but only in close consultation with local consumers.

Urban communities may also opt for a fourth option – improved public sector provision. One of the most successful instances of this is the turnaround in performance at the public sewer and water company in Bogotá, Colombia, Empresa de Acueducto y Alcantarillado de Bogotá (EAAB). In the early 1990s, EAAB was practically bankrupt as a result of the challenges posed by servicing a population that was expanding by 180,000 residents per year. This was made worse by a lack of professional management due to rampant political cronyism (Ronderos 2004: 58). Rejecting advice (and money) from the World Bank, the city of Bogotá decided to strengthen the EAAB with a package of reforms that included reducing subsidies, encouraging conservation, and re-investing profits in expanding infrastructure. Between 1993 and 2001, the percentage of the population of Bogotá that had access to clean drinking water climbed from 78 per cent to 95 per cent (Ronderos 2004: 59). This is an illustration of what has been termed the 'traditional hydraulic paradigm' where near complete public control of water-resources development by the state is favoured on the basis that 'water is a "public" rather than a "tradable" good whose provision is best undertaken as a service rather than a business by the private sector' (Bakker 2002: 771).

If Bogotá is an example of a city that rejected the siren of water privatisation, Cochabamba, Bolivia, is a case of another urban centre that did not, with unfortunate results.

Cochabamba is a city of 800,000, the third largest in the country. It has had a chronic water shortage for years. As is the case in many Latin American cities, Cochabamba has witnessed considerable population growth in recent years, much of it the result of an influx of rural migrants. Most end up in unserviced shanty towns that ring the city. The majority of these *barrios marginales* are not hooked up to the drinking water network, so state subsidies went mainly to industries and middle-class neighbourhoods (Finnegan 2002: 44). In addition to the municipal water system, there are a number of independent, smaller water cooperatives that pump out clean water from single wells.

Since 1985, much of the public infrastructure in Bolivia – railways, the telephone system, the national airline – has been sold, mainly to foreign investors. In 1999, water joined the list when the Bolivian government conducted an auction of the Cochabamba water system as a part of its privatisation programme. The only bid, duly accepted, was from Aguas del Tunari, a consortium controlled at the time by the giant American engineering contractor Bechtel. According to the terms of the contract, Aguas del Tunari would take over the existing municipal water system, rehabilitate it and build new water storage and distribution lines (Lipschutz 2004: 193). The contract would guarantee the company a minimum 15 per cent annual return on its investment, adjusted annually to the consumer price index in the United States (Finnegan 2002: 45).

The deal was problematic for several reasons. First, the contract allowed the company exclusive water rights in the metropolitan district. This meant that the score of cooperatively built small water cooperatives faced expropriation and their wells would be metered.

Second, rates were to be raised, in some cases doubled. Some of this increase was earmarked for rehabilitating a deteriorating water system but other funds were to be used to complete the Misicuni dam, a megaproject that was pronounced uneconomic by the World Bank, but favoured by the mayor, a former real estate developer (p. 47).

In February 2000, a 'water war' began in Cochabamba. A series of demonstrations were held in the central plaza, followed by a four-day general strike. Most of the protest supporters were drawn from the informal sector of pieceworkers, sweatshop employees and street venders, who were available because they had flexible work schedules, as did students from the local university (pp. 49–50). Soon the conflict sharpened, hundreds of injuries occurred and the leaders were arrested. When the protests spread elsewhere in the country, President Hugo Banzer declared a state of siege and ordered mass arrests. Eventually, the company's executives were told that the police could no longer guarantee their safety and fled the city, convincing the national government to cut its losses and revoke the contract. Bechtel through its subsidiary sued the Bolivian government in a World Trade court in Washington for $25 million in profits lost as a result of the cancellation of the contract.

A new national water law was quickly passed that gave legal recognition to traditional communal practices, protected small independent water systems, guaranteed public consultation on rates and gave social needs priority over financial goals (p. 51). Unfortunately, little has been done to improve service and SEMPA, the now reinstated municipal water service, cannot find new capital.

Protests against water privatisation have occurred elsewhere in Bolivia and in other countries. In Panama, popular anger about an attempted privatisation helped cost the President his bid for re-election, while major water privatisations in Lima, Peru, and Rio de Janeiro, Brazil, were cancelled because of popular opposition (p. 53). In Ghana, protests occurred when a privatisation deal accepted by the government in order to receive an IMF loan doubled water rates (Luoma 2004: 56). This forced the World Bank to withdraw from a contract to provide water for the capital city of Accra (Vidal 2005b). A flagship $140 million water privatisation scheme in the coastal city of Dar es Salaam, Tanzania, financed by the World Bank and contracted to the British firm Biwater, recently collapsed amid mutual recriminations. 'Resentment against private water monopolies is growing', the *Guardian* correspondent John Vidal notes, and 'there have been demonstrations in South America, Africa, the Caribbean and Asia' (2005a).

Situations such as these challenge political ecologists to 'attend to how discursive relations – and not just market relations – organize social and ecological change' (Braun and Castree 1998: 16). While the economic inequalities involved are real, as are the protests that arise from them, this is also an ideological conflict and a collision between fundamental values. At the core is a 'discursive move' from the conceptualisation of water as a 'tradable' rather than a 'public' good (Bakker 2002: 770).

While the former may have become a 'dominant global discourse' (Haughton 2002: 806), it has scarcely gone uncontested. Discursive claims on both sides have been registered in a variety of media. Thus, Canadian writer Varda Burstyn (2005) has published a science fiction thriller entitled *Water, Inc.* depicting the struggles in Quebec of a band of eco-revolutionaries against the powers of big business to protect water. Also, the Second World Water Forum represents a deliberate effort by 'a partisan set of individuals from private companies and multilateral agencies to promote privatisation'. (Haughton 2002: 802).

'Tricklers and bumpenaze meters'

No country has witnessed more division over this than South Africa. When apartheid ended in 1994, Nelson Mandela and the African National Congress wrote a clause (Section 27) into the new constitution recognising access to drinking water as a right of citizenship. Their favoured approach, however, was to embrace Thatcherite policies, notably the privatisation of water delivery. This direction was encouraged by the World Bank, which sent its experts to South Africa to help fashion a new economic policy that revolved in part around privatising utilities. In November, 1994, Bank staff drafted the main sections of South Africa's 'Urban Infrastructure Investment Framework', issued four months later under the auspices of the Reconstruction and Development Ministry in President Mandela's office. Within two years, under continuing advice from World Bank advisors, the government added provisions that stipulated a reduced role for the state, fiscal restraint and the promotion of privatisation (Pauw 2005).

The new policy dictated 'full cost recovery' for drinking water provision. This was applied in different ways, with some cities keeping their municipal systems but reducing subsidies, while others signed contracts with French and British water corporations. Throughout the poor black townships of the country, this process became known as 'Water for Profit' (Carty 2003).

On the ground, the water privatisation initiative utilised two technological control devices. In the townships where communal taps were the norm, 'Bumpenaze' meters were installed that could only be accessed through use of a prepaid water card that resembles a phone card. Delinquent individual customers were equipped with a 'trickler', a perforated disc that only permits a small flow of water flow through the tap.

According to Canadian academic and development worker David McDonald, in the decade since this system was put in place as many as 10 million South Africans had their water cut off (Carty 2003). In the community of Ngwelezane in the eastern state of Kwazulu-Natal, where most of the meters on the communal pipes had broken down and many of the individual taps were locked by the local water utility, some people (one survey cited a figure of 11 per cent) began drawing their water from local ponds and streams, polluted with high bacteria counts. This coincided with a cholera outbreak in August 2000, the worst in South Africa's recent history.

As has occurred other places, protests began to occur, ranging from unauthorised water line reconnections to the tyre burning that was common in the black townships during the anti-apartheid era. During the World Summit on Sustainable Development in Johannesburg, 20,000 demonstrators took to the streets, many holding placards saying 'our water is not for sale' (Carty 2003). In the face of these events, governments modified their water policies, although the 'pay for use' policy still applied. In some municipalities, contracts with private companies were cancelled.

5 Social construction of environmental issues and problems

Central to the social construction of environmental issues and problems is the idea that these do not rise and fall according to some fixed, asocial, self-evident set of criteria. Rather, their progress varies in direct response to successful 'claims-making' by a cast of social actors that includes scientists, industrialists, politicians, civil servants, journalists and environmental activists.

Environmental problems are similar in many ways to social problems in general. There are, however, a few notable differences. While social problems frequently cross over from a medical discourse to the arenas of public discourse and action (Rittenhouse 1991: 412), they nevertheless derive much of their rhetorical power from moral rather than factual argument. By contrast, environmental problems such as pesticide poisoning or global warming are tied more directly to scientific findings and claims (Yearley 1992: 117). This is true even in the case of environmental justice claims, which are among the most morally charged indictments of corporate and state polluters. Furthermore, although they are traceable to human agents, environmental problems have a more imposing physical basis than social problems, which are more rooted in personal troubles that become converted into public issues (Mills 1959).

The constructionist interpretation has one primary set of roots in a paradigm shift that transformed the 'sociology of social problems' in the early 1970s.

Constructing social problems

Nearly a quarter of a century ago, the sociology of social problems first began to experience a major conflict with the appearance of a seminal article by Malcolm Spector and John Kitsuse (1973) entitled 'Social problems: a reformulation'. Here, and in a subsequent book (1977), Spector and Kitsuse challenged the 'structural functional' approach to social problems that had theretofore dominated the field. Functionalism, as exemplified by the work of Merton and Nisbet (1971), assumed the existence of social problems (crime, divorce, mental illness) which were the direct products of readily identifiable, distinctive and visible objective conditions. Sociologists were regarded as experts who employ scientific methods to locate and analyse these moral violations and advise policy-makers on how best to cope. In addition, the sociologist's role was to bring to lay audiences an awareness and understanding of worrisome conditions, especially where these were not readily evident (Gusfield 1984: 39).

Spector and Kitsuse argued that social problems are not static conditions but rather 'sequences of events' that develop on the basis of collective definitions. Accordingly, they defined social problems as 'the activities of groups making assertions of grievances and claims to organizations, agencies and institutions about some putative conditions' (1973: 146). From this point of view, the process of claims-making is treated as more important than the task of assessing whether these claims are truly valid or not. For example, rather than document a rising crime rate, the social problems analyst is urged to focus on how this problem is 'generated and sustained by the activities of complaining groups and institutional responses to them' (1973: 158). Since 1973, social constructionism has increasingly moved towards the core of social theorising, generating a critical mass of theoretical and empirical contributions both within the social problems area and across sociology as a whole.

Constructionism as an analytic tool

Best (1989: 250) has noted that constructionism is not only helpful as a theoretical stance but also that it can be useful as an analytic tool. In this regard, he suggests three primary foci for studying social problems from a social constructionist perspective: the claims themselves; the claims-makers; and the claims-making process.

Nature of claims

As initially conceptualised by Spector and Kitsuse, claims were complaints about social conditions which members of a group perceived to be offensive and undesirable. According to Best (1989: 250), there are several key questions to be considered when analysing the content of a claim: What is being said about the problem? How is the problem being typified? What is the rhetoric of claims-making? How are claims presented so as to persuade their audiences? Of these, it is the third question that has generated the most interest among contemporary social problems analysts. Using the example of the 'missing children', e.g. runaways, child-snatched abductions by strangers, Best (1987) analyses the content of social problems claims by focusing on the 'rhetoric' of claims-making. Rhetoric involves the deliberate use of language in order to persuade. Rhetorical statements contain three principal components or categories of statements: grounds, warrants and conclusions.

Grounds or data furnish the basic facts that shape the ensuing policy-making discourse. There are three main types of grounds statements: definitions, examples and numeric estimates. Definitions set the boundaries or domain of the problem and give it an orientation, that is, a guide to how we interpret it. Examples make it easier for public bodies to identify with the people affected by the problem, especially where they are seen as helpless victims. Atrocity tales are one especially effective type of example. By estimating the magnitude of the problem, claims-makers establish its importance, its potential for growth and its range (often of epidemic proportions).

Warrants are justifications for demanding that action be taken. These can include presenting the victim as blameless or innocent, emphasising links with the historical past or linking the claims to basic rights and freedoms. For example, in analysing the professional literature on 'elder abuse', Baumann (1989) identified six primary warrants: (1) the elderly are dependent; (2) the elderly are vulnerable; (3) abuse is life-threatening; (4) the elderly are incompetent; (5) ageing stresses families; (6) elder abuse often indicates other family problems.

Conclusions spell out the action that is needed to alleviate or eradicate a social problem. This frequently entails the formulation of new social control policies by existing bureaucratic institutions or the creation of new agencies to carry out these policies.

Best further proposes two rhetorical themes or tactics which vary according to the nature of the target audience. The *rhetoric of rectitude* (values or morality require that a problem receive attention) is most effective early on in a claims-making campaign when audiences are more polarised, activists are less experienced and the primary demand is for a problem to be viewed in a new way. By contrast, the *rhetoric of rationality* (ratifying a claim will earn the audience some type of concrete benefits) works best at the later stages of social problems construction when claims-makers are more sophisticated, the primary demand is for detailed policy agendas and audiences are more persuadable. Rafter (1992: 27) has added another rhetorical tactic to Best's list: that of archetype formation. *Archetypes* are the templates from which stereotypes are minted and therefore possess considerable persuasive power as part of a claims-making campaign.

A further set of rhetorical strategies in claims-making has been proposed by Ibarra and Kitsuse (1993) who outline a variety of rhetorical idioms, motifs and claims-making styles.[1]

Rhetorical idioms are image clusters that endow claims with moral significance. They include a 'rhetoric of loss' (of innocence, nature, culture, etc.); a 'rhetoric of unreason' that invokes images of manipulation and conspiracy; a 'rhetoric of calamity' (in a world full of deteriorating conditions, epidemic proportions are claimed for a few; for example, AIDS or the greenhouse effect); a 'rhetoric of entitlement' (justice and fair play demand that the condition, or as Ibarra and Kitsuse term it, the 'condition-category', be redressed), and the 'rhetoric of endangerment' (condition-categories pose intolerable risks to one's health or safety).

Rhetorical motifs are recurrent metaphors and other figures of speech (AIDS as a 'plague', the depletion of the ozone layer as a 'ticking time bomb') that highlight some aspect of a social problem and imbue it with a moral significance. Some motifs refer to moral agents, others to practices and still others to magnitudes (Ibarra and Kitsuse 1993: 47).

Claims-making styles refer to the fashioning of a claim so that it is synchronous with the intended audience (public bodies, bureaucrats, etc.). Examples of claims-making styles include a scientific style, a comic style, a theatrical style, a civic style, a legalistic style and a subcultural style. Claims-makers must match the right style to the situation and audience.

Claims-makers

In looking at the identity of claims-makers, Best (1989b: 250) advises that we pose a number of questions. Are claims-makers affiliated to specific organisations, social movements, professions or interest groups? Do they represent their own interests or those of third parties? Are they experienced or novices? (As we have seen, this can influence the choice of rhetorical tactics.)

Many studies that have been undertaken in the social constructionist mode have pointed to the important role played by medical professionals and scientists in constructing social problems claims. Others have noted the importance of policy or issue entrepreneurs – politicians, public interest law firms, civil servants – whose careers are dependent upon creating new opportunities, programmes and sources of funding. Claims-makers may also reside in the mass media, especially since the manufacture of news depends upon journalists, editors and producers constantly finding new trends, fashions and issues.

The cast of claims-makers who combine to promote a social problem can be quite diverse. For example, Kitsuse *et al.* (1984) identify three main categories of claims-makers in the identification of the *kikokushijo* problem in Japan, that is, the educational disadvantage of Japanese schoolchildren whose parents have taken them abroad as part of a corporate or diplomatic posting: officials in prestigious and influential government agencies; informally organised groups of diplomatic and corporate wives; and the '*meta*' – a support group of young adults who have been victims of the *kikokushijo* experience.

It is also important to keep in mind that not all claims-makers are to be found among the grassroots or civil society. For example, it has been suggested that the contemporary 'obesity crisis' has been captained by 'a relatively small group of scientists and doctors, many directly funded by the weight-loss industry, [who] have created an arbitrary and unscientific definition of overweight and obesity' (Oliver 2005, cited in Gibbs 2005: 72).

Claims-making process

Wiener (1981) has depicted the collective definition of social problems as a continually ricocheting interaction among three sub-processes: *animating the problem* (establishing turf rights, developing constituencies, funnelling advice and imparting skills and information); *legitimating the problem* (borrowing expertise and prestige, redefining its scope, e.g. from a moral to a legal question, building respectability, maintaining a separate identity); and *demonstrating the problem* (competing for attention, combining for strength, i.e. forging alliances with other claims-makers, selecting supportive data, convincing opposing ideologists, enlarging the bounds of responsibility). These are overlapping rather than sequential processes which together result in a public arena being built around a social problem.

Hilgartner and Bosk (1988) have identified these *arenas of public discourse* as the prime location for the evaluation of social problem definitions. However, rather than examining the stages of problem development, they propose a model which stresses the competition among potential social problems for attention, legitimacy and societal resources. Claims-makers or 'operatives' are said to deliberately adapt their social problem claims to fit their target environments; for instance, by packaging their claims in a novel, dramatic and succinct form or by framing claims in politically acceptable rhetoric.

Best (1989b: 251) poses a number of useful questions about the claims-making process. Whom did the claims-makers address? Were other claims-makers presenting rival claims? What concerns and interests did the claims-makers' audience bring to the issue, and how did these come to shape the audience's responses to the claims? How did the nature of the claims or the identity of the claims-makers affect the audience's response?

Key tasks/processes in the social construction of environmental problems

In defining environmental problems, bringing them to society's attention and provoking action, claims-makers must engage in a variety of activities. Some of these are centrally concerned with the collective definition of potential problems, others with the collective action necessary to ameliorate them (Cracknell 1993: 4). This is not to say that elements of definition and action do not interweave constantly. Nevertheless, environmental problems do follow a certain temporal order of development as they progress from initial discovery to policy implementation.

In this section of the chapter, I identify three central tasks that characterise the construction of environmental problems. In doing so, I draw upon two prior models: Carolyn Wiener's (1981) three processes through which a public arena is built around a social problem, and William Solesbury's (1976) three tasks which are necessary for an environmental issue to originate, develop and grow powerful within the political system.

As already noted earlier in this chapter, in her book *The Politics of Alcoholism*, Wiener depicted the collective definition of social problems as a continuing ricocheting interaction among three processes: animating, legitimising and demonstrating the problem. These are presented as overlapping rather than sequential processes; that is, they interact with one another rather than operate independently.

Solesbury's scheme is more concerned with the political fate of environmental concerns. He notes the 'continuing change in the agenda of environmental issues' that may be partly accounted for by changes in the state of the environment itself (see Ungar 1992) and partly through changing public views as to which issues are important and which are not. All environmental issues, he states, must pass three separate tests: commanding attention, claiming legitimacy and invoking action. Like Wiener, Solesbury points out that these tasks may be pursued simultaneously in no particular order (Cracknell 1993: 5), although it would presumably be difficult to invoke policy changes before the problem is recognised and legitimised.

In considering the social construction of environmental problems, it is possible to identify three key tasks: assembling, presenting and contesting claims.

Assembling environmental claims

The task of assembling environmental claims concerns the initial discovery and elaboration of an incipient problem. At this stage, it is necessary to engage in a variety of specific activities: naming the problem, distinguishing it from other similar or more encompassing problems, determining the scientific, technical, moral or legal basis of the claim, and gauging who is responsible for taking ameliorative action.

Environmental problems frequently originate in the realm of science. One reason for this is that ordinary people have neither the expertise nor the resources to find new problems. For example, knowledge about the ozone layer is not tied to our everyday experience; it is available only through the use of high technology probes into the atmosphere above the polar regions (Yearley 1992: 116).

Some problems, however, do relate more closely to our life experiences. Concern over toxic wastes frequently begins with local citizens who come to draw a causal link between seeping dump sites and a perceived increase in the neighbourhood incidence of leukaemia, miscarriages, birth defects and other health problems. This is what occurred in Niagara Falls, New York State, where Lois Gibbs and her neighbours were the first to associate their health-related problems with the chemical wastes buried thirty years before in the

Table 2 Key tasks in constructing environmental problems

	Task		
	Assembling	*Presenting*	*Contesting*
Primary activities	discovering the problem naming the problem determining the basis of the claim establishing parameters	commanding attention legitimating the claim	invoking action mobilising support defending ownership
Central forum	science	mass media	politics
Predominant layer of proof	scientific	moral	legal
Predominant scientific role(s)	trend spotter	communicator	applied policy analyst
Potential pitfalls	lack of clarity ambiguity conflicting scientific evidence	low visibility declining novelty	co-optation issue fatigue countervailing claims
Strategies for success	creating an experiential focus streamlining knowledge claims scientific division of labour	linkage to popular issues and causes use of dramatic verbal and visual imagery rhetorical tactics and strategies	networking developing technical expertise opening policy windows

abandoned Love Canal. Those whose jobs or recreational pursuits bring them into close contact with nature on a daily basis (farmers, anglers, wildlife officers) may also be the initial source of claims because they pick up early environmental warning signals such as reproductive problems in livestock or mutations in fish. Acid rain was first launched as a contemporary environmental problem when a fisheries inspector in a remote area of Sweden telephoned researcher Svante Oden with the observation that there appeared to be a link between a rising incidence of fish deaths and an elevation in the acidity of lakes and rivers in the area.

Practical knowledge about the environment often originates from the everyday experience of villagers, small farmers and others in Southern societies. Sir Albert Howard, often regarded as the originator of organic agriculture, derived many of his ideas from consulting with peasant cultivators in India whom he called his 'professors' (Howard 1953: 222) a strategy which was considered revolutionary in the context of British colonial administration. More recently, grassroots activists in Third World countries have emphasised the importance of 'ordinary knowledge' (Lindblom and Cohen 1979) that depends more on keen observation and common sense than on professional techniques. This ordinary knowledge is accumulated within local grassroots networks by breathing air, drinking water, tilling soil, harvesting forest produce and fishing rivers, lakes and oceans (Breyman 1993: 131). In a similar fashion, native (aboriginal) people in Northern societies accumulate firsthand knowledge of the environment that may not be available to non-indigenous observers. For example, it has been suggested[2] that biologists estimating the effect of mega-projects on the ecology of rivers in the Canadian north may overlook the existence of a number of fish species simply because they never bother to ask native residents who know the land intimately (Richardson *et al.* 1993: 87).

In researching the origins of environmental claims, it is important for the researcher to ask where a claim comes from, who owns or manages it, what economic and political interests claims-makers represent and what type of resources they bring to the claims-making process.

In the early US conservation movement, environmental claims were largely traceable to an East Coast élite who utilised a network of 'old boy' ties to secure funding and political action. Enthusiastic amateurs, they dominated the boards of zoos, natural history museums and other public institutions from where they were able to direct campaigns to save redwood trees, migratory birds, the American bison and other endangered species and habitats (Fox 1981). In a similar fashion, the threat to British birds, wildlife sites and other elements of nature was proclaimed in the late nineteenth and early twentieth centuries by a number of conservation groups with élite membership (Evans 1992; Sheail 1976).

By contrast, present day environmental claims-makers are more likely to take the form of professional social movements with paid administrative and research staffs, sophisticated fund-raising programmes and strong, institutionalised links both to legislators and the mass media. Some groups even use door-to-door canvassers who are paid an hourly wage or get to keep a percentage of their solicitations. Campaigns are planned in advance, often in pseudo-military fashion. Grassroots participation is not encouraged beyond 'paper memberships' with control centralised in the hands of a core group of full-time activists.

The process of assembling an environmental claim often involves a rough division of labour. While there are notable exceptions, research scientists are normally handicapped by a combination of scholarly caution, excessive use of technical jargon and inexperience in handling the media. As a result, an important finding may lie fallow for decades until proactively transformed into a claim by entrepreneurial organisations (Greenpeace, Friends of the Earth, Sierra Club) or individuals (Paul Ehrlich, Jeremy Rifkin). Greenpeace's claims-making activity, for example, does not so much flow out of its ability to construct entirely new environmental problems but rather from its genius in selecting, framing and elaborating scientific interpretations which might otherwise have gone unnoticed or been deliberately glossed over (Hansen 1993b: 171). Indeed, the nature of the relationship between the news media and environmental pressure groups such as Greenpeace has become sufficiently institutionalised (Anderson 1993a: 55) that it would be difficult for an emergent problem to penetrate the mass media arena without at least token validation from the latter.

In assembling an environmental problem, not all explanations are created equally. Claims that hinge on difficult to understand concepts such as 'entropy' are far less likely to stick than those that have at their nucleus more readily comprehensible constructs, for example, 'extinction' or 'overpopulation'. Sometimes, the basic outline of a claim only becomes clear in the context of a political, economic or geographic 'crisis'. This was the case in 1973 when concerted action by OPEC (Organisation of Petroleum Exporting Countries), the oil producers' cartel, triggered an energy crisis in industrial nations in the West. Similarly, the abnormally hot US summer of 1988 gave the problem of global warming a visible, experiential focus.

Presenting environmental claims

In presenting an environmental claim, issue entrepreneurs have a dual mandate: they need both to command attention and to legitimate their claim (Solesbury 1976). While not unrelated, these constitute two quite separate tasks.

As Hilgartner and Bosk's (1988) model emphasises, the arenas through which social problems become defined and conveyed to the public are highly competitive. To command attention, a potential environmental problem must be seen to be novel, important and understandable – the same values which characterise news selection in general (Gans 1979).

One effective way of commanding attention is through the claimants' use of evocative verbal and visual imagery. Thus the extreme thinning of the ozone layer became much more saleable as an environmental problem when depicted as an expanding 'hole'; American children's entertainer Bill Shontz has even recorded a hit song entitled *Hole in the Ozone*. Similarly, the effects of acid rain were successfully dramatised when German environmentalists began to use the term *Waldsterben* (forest die-back). More recently, Larson *et al.* (2005) have demonstrated the prevalence of militaristic metaphors (attack, destroy, wipe out, contain, counteroffensive, full-scale war) in the media reporting of three contested areas of science–society discourse (invasive species, foot-and-mouth disease and SARS (Severe Acute Respiratory Syndrome). Visual language can be especially powerful in carrying out this task. For example, technical data on the size of seal herds and codfish stocks instantly lost relevance when Brian Davies and other activists released photos to the media of baby seal pups being clubbed to death on the ice floes of Labrador.

It is not unusual, however, for these visual images to be streamlined so as to underline a central image. Mazur and Lee (1993: 711) give several striking examples of this. The NASA satellite pictures of the ozone hole over the Antarctic, which became a 'logo' of the problem, transformed continuous gradations in real ozone concentration into an ordinal scale that is colour-coded, conveying the erroneous impression that a discrete, identifiable hole could actually be located in the atmosphere over the South Pole. In August 1988, a *New York Times* article on rainforest destruction was accompanied by a stunning satellite photograph of the burning Amazon that was created by Alberto Setzer of the Brazilian Institute of Space Research. The photograph showed what appeared to be nearly 100,000 fires; however, it was really a composite of many separate pictures and included fires in areas of secondary forest growth as well as virgin rainforest.

Environmental issues may be forced into prominence when exemplified by particular incidents or events, for example, the nuclear accidents at Chernobyl and Three Mile Island, the Bhopal chemical disaster, the wreck of the oil tankers *Torrey Canyon* and *Exxon Valdez*. Dramatic events like these are important because they assist political identification of the nature of an issue, the situations out of which it arises, the causes and effects, the identity of the activities and the groups in the community which are involved with the issue (Solesbury1976: 384–5).

Staggenborg (1993) has identified six major types of 'critical events' that affect social movements such as the environmental movement. Large-scale socio-economic and political events such as wars, depressions and national elections influence the opportunities for collective action by altering perceptions of grievances and threats; for example, the 1980 election of US President Ronald Reagan led to increased memberships in environmental groups[3] since it raised the spectre of a free enterprise run rampant in national parks and other wilderness settings. National disasters and epidemics can represent a turning point in the movement, highlighting grievances and bringing about movement growth. Similarly, industrial and nuclear accidents can be potentially useful to the movement by laying bare policies and features of the power structure that are normally hidden; for example, the power of the oil companies in the Santa Barbara oil spill (Molotch 1970). Critical encounters involve face-to-face interaction between authorities and other movement actors focusing attention on movement issues. Strategic initiatives are events created by deliberate actions taken by supporters or opponents to advance movement or counter-movement goals. The staged events that are characteristic of Greenpeace campaigns are examples of this, as is the publication of polemical books such as Paul Ehrlich's *The Population Bomb* and Jeremy Rifkin's *Beyond Beef*. Finally, policy outcomes are official responses to collective action by a movement or counter-movement – critical junctures at which movements are forced to renegotiate their strategies, tactics and goals as a result of changes in the political environment. The decision by the Roosevelt administration in 1914 to begin construction of the Hetch Hetchy Dam in Yosemite National Park in order to provide water for a pipeline to San Francisco was such a decision, in that it destroyed any possibility of a further alliance between the resource conservationists as represented by Gifford Pinchot and the preservationists led by John Muir.

Staggenborg's discussion is directed primarily towards the issues of social movement mobilisation and strategies, but her typology of events is relevant to the presentation of environmental claims insofar as environmental organisations often represent the primary claims-makers at this stage of the construction of environmental problems.

Of course, not all critical events are guaranteed to generate a high profile problem. According to Enloe (1975: 21), an event provokes an environmental issue when it (1) stimulates media attention; (2) involves some arm of the government; (3) demands governmental decision; (4) is not written off by the public as a freak, one-time occurrence; and (5) relates to the personal interests of a significant number of citizens. These criteria are partly a function of the incident itself but also depend on the successful exploitation of the event by environmental promoters.

In presenting environmental claims, movement leaders engage in what Snow *et al.* (1986) call the process of 'frame alignment'; i.e. environmental groups tap into and manipulate existing public concerns and perceptions in order to broaden their appeal. For example, Greenpeace primarily chooses topics and organises campaigns in areas that can lend themselves to the widest public resonance (Eyerman and Jamison 1989: 112) while avoiding those which are more divisive. In a similar fashion, environmental movement opponents attempt to appeal to a wider public by linking new technologies or programmes to popular issues and causes. Thus the biotechnology industry has tried to foster a public image of an incremental and benign technology that is useful in promoting economic development (Plein 1991). Commanding attention is not, however, sufficient to get a new issue on the agenda for public debate (Solebury 1976: 387). Rather, emergent environmental problems must be legitimated in multiple arenas – the media, government, science and the public.

One way to achieve this legitimacy is through the use of the rhetorical tactics and strategies cited by Best (1987) and Ibarra and Kitsuse (1993). Rather than follow a chronological order, as Best suggests, environmental rhetoric has become increasingly polarised. Ecofeminists, deep ecologists and other purveyors of what Dryzek (2005) calls 'green radicalism' have tended to adopt a 'rhetoric of rectitude' which justifies consideration of environmental problems on strictly moral grounds. By contrast, environmental pragmatists, who advocate sundry versions of the 'sustainable development' paradigm, tend towards a rhetoric of rationality. Green business, for example, is based on the premise that environmentalism can be both socially useful and profitable.

This cleavage can be illustrated with reference to the loss of tropical rainforests in Brazil, Malaysia and Indonesia. Pragmatists argue that the loss of these rainforests is a serious problem because it leads to the extinction of rare indigenous insects, plants and animals that are invaluable to pharmaceutical companies as sources of new wonder drugs. Environmental purists, on the other hand, base their claims on a rhetoric that stresses the inherent spiritual value of these endangered habitats.[4]

Environmental claims can also be legitimated when their sponsors become legitimate and authoritative sources of information. Hansen (1993b) has demonstrated that Greenpeace has achieved this kind of sustained success as a claims-maker in a number of ways: by acting as a conduit for the dissemination of new scientific developments between the research community and the media; by becoming a 'shorthand signifier' for everything environmental – environmental caring, green lifestyles, environmentally conscious attitudes – and by producing knowledge and information which can be used strategically in public arena debates (see Eyerman and Jamison 1989).

It is sometimes possible to pinpoint an event which constitutes the turning point for an environmental problem and when it breaks through into the zone of legitimacy. With regard to global warming, this occurred at US Senate hearings in 1988 when Dr James Hansen made the claim that he was 99 per cent sure that the warming of the 1980s was not due to chance but rather to global warming. In the case of ozone depletion, the key event was a 1988 NASA/NOAA report providing hard evidence for the first time implicating CFCs (chlorofluorocarbons) in ozone layer depletion. With pulp mill dioxins, it was the 1987 release of the '5 Mill Study' showing that traces of this toxic chemical had been detected in various household paper products and the subsequent front-page story in the *New York Times* that launched this problem in the United States, and, later, in Canada (Harrison and Hoberg 1991).

Yet scientific findings and testimony by themselves are not always sufficient to push an environmental problem past the break point of legitimacy. In the case of global warming, Dr Hansen's earlier Senate testimony in 1986, where he predicted that significant global warming might be felt within five to fifteen years, did not attract comparable coverage or concern. This only occurred two years later when there had been a significant shift in media practices and public attention (Ungar 1992: 492). Similarly, Molina and Rowland's 1974 publication in the journal *Nature* of their theory that CFCs were destroying the ozone layer at first only brought limited coverage in the California press. It was only later on when the issue became linked to claims that other gases from aerosol cans, notably vinyl chloride, were linked to skin cancer, that their data were given wide attention and media legitimacy (Mazur and Lee 1993: 686).

Contesting environmental claims

Even if an emergent environmental claim manages to transcend the threshold of legitimacy, this does not automatically ensure that an ameliorative action will be taken. As Gould *et al.* (1993: 229) have noted, one can interpret environmental protection history from the position that environmental movements have been far more successful in getting listed on the broad political agenda than in getting their policies within this agenda, especially where these policies might require the reallocation of resources away from large-scale capital interests and state bureaucratic actors.

Solesbury (1976: 392–5) has noted a number of factors that can contribute to an issue being lost at the point of decision or action. Major external constraints such as the onset of a national economic crisis may lead to a problem being postponed, then altogether abandoned. A problem may be transformed into a less threatening political issue. Opponents within government bureaucracies may use a number of tactics – postponing discussion, referring an item back for further research or amendment – which ensure that a problem will not immediately be acted upon.

As a consequence, invoking action on an environmental claim requires an ongoing contestation by claims-makers seeking to effect legal and political change. While scientific support and media attention continue to constitute an important part of the claim package, the problem is principally contested within the arena of politics. Contesting an environmental problem within the political policy stream is a fine art, given the cross pressures which legislators face.

Environmental entrepreneurs must skilfully guide their proposals through a log jam of vested and often conflicting political interest groups, each of which is capable of stalling or sinking the proposals. As Walker has noted:

> Public [environmental] policies seldom result from a rational process in which problems are precisely identified and then carefully matched with optimal solutions. Most policies emerge haltingly and piecemeal from a complicated series of bargains and compromises that reflect the biases, goals and enhancement needs of established agencies, professional communities and ambitious political entrepreneurs.
>
> (1981: 90)

Kingdon (1984) observes that policy proposals that survive in this political jungle usually satisfy several basic criteria.

First, legislators must be convinced that a proposal is technically feasible; that is, if enacted, the idea will work. This may not prove to be the case in hindsight; for example, the Endangered Species Act in the United States has worked out much less perfectly in its implementation than on paper. Nevertheless, a proposal must at least initially appear to be scientifically sound and politically administrable.

Second, a proposal that survives in the political community must be compatible with the values of policy-makers. Since most bureaucrats and politicians do not hold ecocentric views, this means that solutions which reflect the New Ecological Paradigm are not likely to get very far unless there is a generally perceived crisis. Instead, environmental solutions that appear, on the surface, to be neutral stand a better chance of being accepted than those that seem ideologically tinged. Furthermore, problems that are framed in utilitarian terms often go further than those that are not. This means that arguments made with financial expediency in mind – figures and statistics translated into 'bottom-line' dollars (pounds/euros) – are more likely to resonate than those that are presented solely on the basis of moral justifications (Hunt *et al.* 1994: 200–1).

Environmental policy is by no means a perfectly predictable and consistent enterprise. For example, Milton (1991) has suggested that the British government routinely adopts a contradictory approach to the environment. On domestic pollution issues it adopts a rigid, hierarchical position that tends to retard change. This has been quite evident in, for example, the British response to the acid rain problem. By contrast, on international environmental problems such as global warming, the UK has adopted a more 'entrepreneurial' approach. On wildlife and conservation issues an approach that constitutes a mixture of the hierarchical and the entrepreneurial is favoured. Sometimes, an issue will rise in the policy agenda for totally unexpected reasons. This occurred with the greenhouse effect which initially achieved the stamp of seriousness not in terms of a long-range threat to the world climate but in relation to what was basically a side issue: the environmental implications of the large-scale deployment of the supersonic transport airplane (SST) in the early 1970s (Hart and Victor 1993: 663–4).

Thus successfully contesting an environmental claim in the political arena requires a unique blend of knowledge, timing and luck. This process is often event-driven with a disaster such as the Three Mile Island nuclear accident opening up 'political windows' (Kingdon 1984: 213) that would otherwise remain closed. This is not to say that agenda-setting and legislative action are totally random but that the process is highly contingent upon a number of internal and external factors, many of which are not linked to the obvious merits of the case.

At the same time, there may also be a contest for 'ownership' of an environmental problem. This can be particularly rancorous where one of the contesting parties is drawn from the ranks of those directly victimised by a problem. There are many examples of this in the social problems field ranging from 'deviance liberation movements' such as the American prostitutes' rights campaign (Jenness 1993; Weitzer 1991) to victims' rights groups; for example, those formed by breast cancer patients. This is less common with environmental problems, which generally have a more diffused impact. One significant example, however, is the dispute over the issue of who owns 'biodiversity' both as a resource and as an environmental problem (see Chapter 9). This struggle pits a coalition of small farmers, ecological activists and others in Third World countries against the conservation establishment: biologists, bureaucrats from non-governmental organisations and government ministries dealing with trade and environmental issues.

Hawkins (1993) has identified three ideal-type paradigms that occupy the increasingly contested discourse over global environmental futures. The prevailing 'global managerialist paradigm' advocates the detection and solution of problems in the globalised commons by an existing configuration of nation states and international organisations buttressed by scientific experts and professional environmentalists within international NGOs (non-governmental organisations). This approach downplays local perceptions and definitions of problems, and on occasion may even blame poor people in Third World nations for causing environmental degradation. The 'redistributive development paradigm' recognises the need for greater equity in matters pertaining to development and the environment in Southern countries. It proposes that such inequities can be redressed through a number of innovative measures such as the Green Fund within the World Bank or debt-for-nature swaps. The 'new international sustainability order paradigm' calls for a fundamental restructuring of the world order such that Third World nations claim a more direct voice in establishing a balance between economic and social sustainability.

Hawkins depicts the construction of international environmentalism as reflecting an ongoing struggle among supporters of these three paradigms. The dispute over the ownership of biodiversity is one manifestation of this; the conflict over global climate change is another. Even the language used in defining this contested ground is itself socially constructed. For example, countries of the North have adopted a globalised language to describe the situation in Southern nations in which 'our' environmental problems (climate change, ozone depletion) are caused by 'their' development problems (forest loss, overpopulation), a situation which is solvable only by embracing 'sustainable development' strategies (Redclift and Woodgate 1994: 64–5). At present, the first two paradigms still predominate but the new international sustainability order paradigm appears to be making some significant inroads.

Audiences for environmental claims

In addition to the skill of claims-makers and the severity of the underlying condition itself, the success of a putative environmental claim may also be tied to the magnitude of audiences that are mobilised around that claim. That is, a groundswell of audience support not only marks the rising of a problem but also can constitute a valuable resource in the effort to capture political attention.

For sociologists, the problem is how reliably to gauge the size and influence of audiences. As Ungar (1994: 298) has pointed out, the potential for environmental claims-makers to use public opinion as a resource is paradoxically both enhanced and limited by present polling procedures. That is, public polling today rarely maps support for contested positions, opting instead for broad measures of environmental concern such as the 'New Environmental Paradigm Scale' developed by Riley Dunlap and his colleagues. This produces such a vague barometer of public opinion that virtually any group on the 'pro-environmental' side can claim to represent it but, at the same time, it makes it difficult to gauge specific reactions to specific issues. Alternatively, one can look to other indicators of public support – recycling behaviour, green consumerism, participation in environmental events and mobilisations – but these too are imperfect measures of opinion.

Nevertheless, the tide of public opinion can sweep a claim upwards on to the policy agenda, sometimes in a dramatic fashion. In the 'Alar' controversy in the United States, for example, public fears about toxins translated into a short-run consumer boycott of apples, even though the risk-supporting data were later found to be less reliable than was originally thought. Similarly, public concern about 'mad cow disease' in Britain has been sufficiently grave for governments to act in a precautionary manner not always so evident in the case of other potential risks.

Of course, not all environmental claims succeed in raising the red flag for concerned audiences. Some claims are perceived as being too extreme, too misanthropic or too complex. Others run up against powerful counter-claims. Some fail because the requisite preventive or mitigative response mandates too great a lifestyle sacrifice.

In considering why some environmental claims capture the public eye and others do not, it may be helpful to look to the field of advertising research. In a large-scale comparative study in the 1990s which examined the attitudes of 30,000 consumers in 21 countries, the New York advertising agency Young & Rubicon came up with a marketing model, the 'Brand Asset Valuator', which isolates four key factors that predict how well a specific product will do in the marketplace: uniqueness, relevance, stature and familiarity (Scotland 1994).

In the case of environmental claims, uniqueness or *distinctiveness* refers to the extent to which the public perceives a problem as separate from others of a similar nature. For example, acid rain claims-makers were successful in distinguishing this condition from the more inclusive category of air pollution. Rhetorical strategies are important here in creating distinctive labels for emerging problems as well as devising symbolic codes that can be attached to a claim in order to confer a distinctive identity.

Relevance refers to the degree to which a particular environmental problem matters to the ordinary citizen. This is not always easy to demonstrate, even when the problem is occurring in people's own backyards. It is especially difficult in the case of global environmental problems which have their origins far away in distant parts of the world. Thus extended drought conditions in the poor African nations are of little relevance in the South-western United States, yet regional water shortages which require that local citizens stop watering their lawns and filling their swimming pools are quite meaningful.

Stature denotes how highly a consumer thinks and feels about a particular brand. In the case of the environment, this refers to the attitudes of the public towards the place or people or species under threat. It is no accident that the wildlife protection movement first mobilised in the nineteenth century over the danger posed to our much-loved songbirds by hunters and by the millinery trade. Similarly, national parks and monuments – Yellowstone Park in the United States, the Lake District in Britain, Great Barrier Reef in Australia – have considerable symbolic stature which comes into play if these places are imperilled. By contrast, low-income black and Hispanic communities in the American South that face serious threats from toxic polluters have long been accorded low stature, especially by middle-class audiences.

Finally, *familiarity* refers to how well-known a particular problem is to an audience. The media play an especially important role here in educating us about environments, species or places that may have been beyond our realm of personal experience. For example, in 1992 it was announced that scientists in Central Vietnam had discovered the *sao la*, a goat-like mammal previously unknown to the outside world. Almost overnight, the *sao la* became a media superstar as a result of a media frenzy whipped up by scientists, environmentalists and the press.[5] Celebrated on the pages of *National Geographic* and *People* magazines, it became 'the zoological equivalent of finding a new planet' (Shenon 1994). In some cases, environmental activists may undertake collective action in order to familiarise audiences with a claim. For example, the clear-cutting practices in the old growth forests in British Columbia became widely known in Europe and America, in part because of the extensive media coverage of protests by environmental activists on the logging roads and on the steps of the provincial legislature. Rather than enhancing the stature of a claim, however, familiarity may ultimately produce issue fatigue on the part of the general public, especially if new developments are not forthcoming. This is the case even if a problem is both distinctive and relevant. Indeed, audiences have an inherent sense of fair play that dictates that activities such as unrelenting 'polluter bashing' are unacceptable, even if the criticism is well deserved.

Successful environmental claims, then, must possess elements of vitality and stature that ensure that they will not perish in a sea of disinterest or irrelevance.

Necessary factors for the successful construction of an environmental problem

It is possible to identify six factors that are necessary for the successful construction of an environmental problem. These are as follows.

First, an environmental problem must have scientific authority for and validation of its claims. Science may well be an 'unreliable friend' to the environmental movement as Yearley (1992) has suggested, but nevertheless it is virtually impossible for an environmental condition to be successfully transformed into a problem without a confirming body of data which comes from the physical or life sciences. This is especially so with the newer global environmental problems, whose very existence hinges on a novel scientific construction (see the discussion of biodiversity loss in Chapter 9).

Second, it is crucial to have one or more scientific 'popularisers' who can transform what would otherwise remain a fascinating but esoteric piece of research into a proactive environmental claim. In some cases (Edward Wilson, Paul Ehrlich, Barry Commoner), the popularisers may themselves be employed as scientists; in others (e.g. Jonathan Porritt, Jeremy Rifkin) they are activist authors whose knowledge of science comes secondhand. Whatever their background, these popularisers assume the role of entrepreneurs, reframing and packaging claims so that they appeal to editors, journalists, political leaders and other opinion-makers.

Third, a prospective environmental problem must receive media attention in which the relevant claim is 'framed' as both real and important. This has been the case for most contemporary problems, for example, ozone depletion, biodiversity loss, rainforest destruction, global warming. By contrast, other significant environmental problems fail to make the public agenda because they are not considered especially newsworthy. For example, in many Canadian cities lack of treatment of urban sewage is endemic but this has received scant coverage compared to other pollution problems. As the executive director of the Sierra Legal Defense Fund once pointed out, a volume of sewage equivalent to thirty-two oil-tankers the size of the *Exxon Valdez* is dumped each day into local rivers or bays, yet it is done out of the sight of the public with virtually no attention from the media (Westell 1994).

Fourth, a potential environmental problem must be dramatised in highly symbolic and visual terms. Ozone depletion was not a candidate for widespread public concern until the decline in concentration was graphically depicted as a hole over the Antarctic. The wanton practices of the major forestry companies only became a matter for international outrage when Greenpeace and other environmental groups began to exhibit dramatic photographs of the 'clear-cuts' on Vancouver Island while labelling the area the 'Brazil of the North'.

Images such as this provide a kind of cognitive short cut compressing a complex argument into one that is easily comprehensible and ethically stimulating.

Fifth, there must be visible economic incentives for taking action on an environmental problem. For example, the case for acting boldly to stop biodiversity loss was levered on the argument that the tropical rainforests contained an untapped wealth of pharmaceuticals that would disappear forever if nothing was done. At the same time, environmental claims that carry positive, economic incentives for one group may also involve costs for others, thus provoking sharp opposition.

Necessary factors for the successful construction of an environmental problem

- Scientific authority for and validation of claims
- Existence of 'popularisers' who can bridge environmentalism and science
- Media attention in which the problem is 'framed' as novel and important
- Dramatisation of the problem in symbolic and visual terms
- Economic incentives for taking positive action
- Recruitment of an institutional sponsor who can ensure both legitimacy and continuity

Finally, for a prospective environmental problem to be fully and successfully contested, there should be an institutional sponsor who can ensure both legitimacy and continuity. This is especially important once a problem has made the policy agenda and legislation is sought. Internationally, this can be seen in the important role played by agencies and NGOs associated with the United Nations.

6 Media and environmental communication

In moving environmental problems from conditions to issues to policy concerns, media visibility is crucial. Without media coverage the odds are low that an erstwhile problem will either enter into the arena of public discourse or become part of the political process. For example, it is unlikely that many of the lay public would have become aware of 'mad cow disease' or the purported dangers of genetically modified (GM) foods if it were not for media reportage (Lupton 2004: 187). Indeed, most of us depend on the media to help make sense of the bewildering daily deluge of information about environmental risks, technologies and initiatives.

While the traditional news media are important here (and are the focus of this chapter), there is also an extensive array of other media sources, from documentary television shows on nature and the environment to motion pictures to Internet websites. For example, MTV in the United States broadcasts 'Trippin' a conservation series directed at teenagers. Co-produced by film actress Cameron Diaz, the series presents endangered animals in their natural habitats. Episodes include Diaz in Tanzania and Nepal, gangsta-rapper DMX in the Yellowstone outback and professional surfer Kelly Slater on the Costa Rican coral reefs (Martel 2005).

Indeed, the news media's role as an agent of environmental education and agenda setting is both important and complex. As Schoenfeld *et al.* (1979) have demonstrated, the daily press in the United States was initially slow in grasping the basic substance and style of environmentalism, leaving it to issue entrepreneurs in colleges and universities, government and public interest groups to mobilise concern outside of the media net. In local environmental conflicts, media claims are often viewed sceptically, refracted as they are through the prism of residents' own practical everyday experiences and knowledge (Burgess and Harrison 1993). Rather than actively sparking a response to environmental problems, the media often seem to be a millstone weighing down public discussion of environmental topics in a technical-bureaucratic discourse that excludes interest groups and non-official claims-makers (Corbett 1993: 82).

In this chapter, I will assess the news media's conflicting role in socially constructing environmental issues and problems. Of particular concern is the extent to which the portrait of the environment presented by mainstream journalists represents a critique of the paradigm of technological progress as opposed to simply an extension of the existing corpus of disaster stories. First, however, it is necessary briefly to outline the general process through which the media 'manufacture' news and endow issues and events with symbolic meaning.

Manufacturing news

For many years, mass communication researchers largely took for granted the existence of 'objective' facts and events that could be verified, exclusive of whether or not they were actually covered by the media. Thus floods and hurricanes, political victories and resignations, medical miracles and foreign wars were all thought to have a certifiable existence of their own beyond the newsroom. Reporters, editors, producers and other 'newsworkers' might, on occasion, distort or selectively omit certain happenings but this did not mean that they were not real (Fishman 1980: 13).

In the 1970s, this approach gave way to a very different model, in which events become news only when transformed by the newswork process and not because of their objective characteristics (Altheide 1976: 173). News is conceptualised here as a 'constructed reality' in which journalists define and redefine social meanings as part of their everyday working routine (Tuchman *et al.* 1978). Newsmaking, in turn, is treated as a collaborative process in which journalists and their sources negotiate stories.

Organisational routines and constraints

While the construction of news may be influenced by cultural or political factors, it is generally seen as the result of inescapable organisational routines within the newsroom itself. Schlesinger (1978) observes that rather than being a form of 'recurring accident', news is the product of a fixed system of work whose goal is to impose a sense of order and predictability upon the chaos of multiple, often unrelated events and issues. In his observational study of BBC news, he found that the backbone of each day's newscasts was a 'routine agenda of predictable stories': labour negotiations, parliamentary business, activities of the Royal Family, sport scores, etc. In a similar fashion, Fishman (1980) observed that, rather than dig for information, reporters at a California daily newspaper opted instead for a diet of routine news derived from a mix of scheduled events (press conferences, courtroom trials) and pre-formulated accounts of events (arrest records, press releases); these items were crucial in helping them to meet deadlines and story quotas.

In addition to mandating that news be planned, time also acts as a constraint upon the final product itself. This has the effect of rendering news reports 'incomprehensible rather than comprehensive' (Clarke 1981: 43). In particular, action clips that fit more easily into existing formats, especially television news, are favoured over longer, more nuanced stories that deal with underlying causes and conditions.

Furthermore, by consistently failing to ask the question 'why', the news process 'contributes to *decontextualising* or removing an event from the context in which it occurs in order to *recontextualise* within news formats' (Altheide 1976: 179). This tendency is further encouraged by the use of news 'angles' – frameworks around which a particular content is moulded in order to tell a story. The use of news angles is pervasive in journalism and plays a significant role in determining not only the 'spin' put on a story but also whether a story is suitable in the first place for broadcasting or publication.

Media constructionists have also noted the importance of news sources in shaping story content. Reporters usually stick to a shortlist of trusted source contacts who, on the basis of past experience, can be counted on to be both articulate and reliable. In fact, it is not unknown for source contact lists to be passed down from one reporter to the next. Trusted sources come from various walks of life but they are usually people who function in official roles: politicians, the heads of government agencies, scientists and other experts. Even where the media solicit comment from opponents of the status quo, news sources are invariably drawn from the executive of major social movement organisations such as Greenpeace and Friends of the Earth.

In their study of the 1969 oil spill in Santa Barbara, California, Molotch and Lester (1975) found that powerful figures and organisations with routine access to the media (the President, federal officials, oil company representatives) were far more likely to function as news sources than were conservationists and local officials. These sources exercise considerable social and political power by providing a pre-packaged, self-serving, socially constructed interpretation of a given set of events or circumstances – an interpretation that is readily adopted by journalists who rarely have the time or the specialised knowledge needed to flesh out their own news angle (Smith 1992: 28).

Media discourse

In recent years, media constructionists have looked beyond the social organisation of the newsroom and focused on the process by which journalists and other cultural entrepreneurs develop and crystallise meaning in public discourse (Gamson and Modigliani 1989: 2). This approach takes as its central concern the decoding of media texts – the visual imagery, sound and language produced in the social construction of news and forms of public communication (Gamson *et al.* 1992: 381).

The key element here is that of *media frames*, a concept adapted by several media sociologists in the late 1970s and early 1980s (Gitlin 1980; Tuchman 1978) from Erving Goffman's work on small-group interaction. Frames, like news angles, are organising devices that help both the journalist and the public make sense of issues and events and thereby inject them with meaning. In short, they furnish an answer to the question 'What is it that's going on here?' (Benford 1993: 678). When expressed over time, frames are known as 'story lines' (Gamson and Wolfsfeld 1993: 118).

Even when the details of an event are not disputed, the event can be framed in a number of different ways. For example, the 1993 murder of Liverpool toddler James Bulger by two 10-year-old boys was variously framed by the press as a new low in the continuing economic and moral decline of England, the turning point in the campaign against 'video nasties' (one boy's father had reportedly rented the movie *Child's Play 3* just before the crime), a cautionary tale for harried parents with youngsters in tow and an example of the linkage between school truancy and juvenile crime. Both claims-makers and their opponents routinely compete to promote their favoured frames to journalists as well as to potential supporters. At the same time, newsworkers forge their own frames largely for reasons of efficiency and story suitability. Gamson and Wolfsfeld (1993) depict the interaction between movements and the media as a subtle 'contest over meaning' in which activists attempt to 'sell' their preferred images, arguments and story lines to journalists and editors who, more often than not, prefer to maintain and reproduce the dominant mainstream frames and cultural codes. In the Nicaraguan conflict of the 1980s, for example, peace activists attempted to counter the official frame that the American-sponsored Contras were waging a struggle against Communist expansion by promoting a 'human costs of war are too high' frame (Ryan 1991).

Finally, as Gamson *et al.* (1992: 384) point out, it is wrong to assume that news consumers (readers, audiences) passively accept media frames as they are; they too may decode media images in different ways utilising varying frameworks of interpretation (Corner and Richardson 1993).

Media discourse, therefore, takes the form of a 'symbolic contest' in which competing sponsors of different frames measure their success by gauging how well their preferred meanings and interpretations are doing in various media arenas (Gamson *et al.* 1992: 385).

The process of framing is in many ways comparable to the rhetoric of claims-making in social problem construction (see Chapter 5). Gamson and Modigliani (1989: 3–4) distinguish five *framing devices*: metaphors, exemplars (i.e. historical examples from which lessons are drawn), catchphrases, depictions and visual images; and three *reasoning devices*: roots (a causal analysis), consequences (i.e. a particular type of effect) and appeals to principle (a set of moral claims) which function as a kind of symbolic shorthand in telegraphing the core meaning of a frame.

Furthermore, they introduce the concept of *media packages*. Media packages help to organise these framing devices in cases of complex policy issues such as the use of nuclear power. In analysing television news coverage, news magazine accounts, editorial cartoons and syndicated opinion columns on nuclear power from 1945 to the present day, Gamson and Modigliani isolate seven different interpretive packages: progress; energy independence; the devil's bargain; runaway; public accountability; not cost-effective; and soft paths. As the titles suggest, each package is represented by 'a deft metaphor, catchphrase or other symbolic device' (1989: 3).

Mass media and environmental news coverage

As Schoenfeld *et al.* (1979: 42–3) have demonstrated, prior to 1969 the daily press in the United States had considerable difficulty recognising environmentalism as a topic separate from that of conservation. Conservation was a reasonably well-understood and respectable concern, having been around since the 1880s. It had a known constituency, its own legislative acts and administrative bureaux and even its own universally recognised symbol – Smokey the Bear. By contrast, the central tenet of environmentalism, i.e. that everything is connected to everything else, seemed difficult to grasp in journalistic terms. Similarly, in Britain, the preservation of the countryside, the national heritage and rare species of fauna and flora were all widely accepted as legitimate activities which cut across class lines, but few journalists readily connected them with air pollution, oil spills and other contemporary environmental problems.

During the late 1960s and early 1970s, media coverage of the environment rose dramatically and, for the first time, environmental issues were seen by journalists in both Britain and America as a major category of news (Lacey and Longman 1993; Parlour and Schatzow 1978). Newsworkers began to perceive individual difficulties such as traffic problems or pollution incidents as part of a more general problem of 'the environment' (Brookes *et al.* 1976; Lowe and Morrison 1984).

There are several key events that may be cited in order to explain this upswing in media awareness and understanding of environmentalists' claims. Schoenfeld *et al.* (1979: 43, citing Roth 1978) argue that the most effective environmental message of the century was totally inadvertent: the 1969 view from the moon of a fragile, finite 'Spaceship Earth'.[1] This provided a powerful metaphor with which to frame the environmental message. Similarly, Earth Day 1970 acted as a news 'peg' for a variety of otherwise disparate news stories on environmentally related subjects, earning extensive coverage both nationally and in many local American communities (Morrison *et al.* 1972)

After 1970, however, media coverage of the environment began to fall off (Parlour and Schatzow 1978), although it recovered briefly during the energy crisis of 1973–4. When stories did appear they were most likely to be event-related and problem-specific. In their examination of article headlines in the *Canadian Newspaper Index*, Einseidal and Coughlan (1993: 140) observed that environmental items were located under a series of disparate and seemingly unconnected problem categories: air pollution; water pollution; waste management; and wildlife conservation. Similarly, Hansen (1993a: xvi) notes the tendency of the media to define the environment 'largely in terms of anything nuclear (nuclear power, nuclear radiation, nuclear waste, nuclear weapons), in terms of pollution and in terms of conservation/protection of endangered species'. Rarely were the global aspects of environmental problems highlighted during this period. Even more unusual was the appearance of stories on environmental problems in Third World countries.

This pattern appears to have changed somewhat in the 1980s and 1990s. Einseidal and Coughlan note that towards the end of 1983 new descriptor terms began to appear in Canadian newspaper headlines; for example, 'global catastrophe', 'environmental order' and 'environmental ethics'. In contrast to the conservation focus, stories were vested with a more global character, encompassing attributes that included 'holism and interdependence and the finiteness of resources' (1993: 141). They also note an increasing urgency and seriousness in the coverage of environmental issues by the Canadian press as indicated by the appearance of a collection of 'war and dominance' metaphors: survival, defeat, battles, crusades. Topic headings were found to be more specific, covering such areas as 'eco-tourism', 'environmental law' and 'eco-feminism'. In a similar fashion, Howenstine (1987) detects a transformation in environmental reporting in major US periodicals from 1970 to 1982 towards a greater complexity of coverage. In addition, he found a shift across time to relatively fewer articles on the degradation and protection of the natural environment and more on and economic and development issues.[2]

However, perception that coverage is deepening may have been overly optimistic. Lacey and Longman (1993) note the rise of a 'show business' and commercial approach to environmental issues in the British media during the 1980s and argue that the improvements in environmental reporting are only evident if a narrow definition of environmental issues is utilised. In particular, an artificial separation is created between the environment and development issues in line with a predominant editorial and political bias. For example, coverage of famine in East Africa has been long on shock tactics but short on political insight, especially in the case of drought in Sudan, a country whose political regime is considered unacceptable by Western policy brokers. Furthermore, the reporting of such stories is cyclical and usually in step with their ascendancy on the political agenda (Anderson 1993b: 55).

Production of environmental news

To a large extent, media coverage of environmental issues is constrained and shaped by the same production constraints that govern newswork in general. Earlier this chapter, we discussed some of the most significant of these: limited production periods; story lengths; and limited sources. Clarke (1992) has grouped these production constraints into two general categories: short-term logistical and technological constraints, and long-term and occupational constraints that are embedded in the news process itself.

Short-term pressures of time have meant that environmental issues and problems have often been framed by journalists within an event orientation. As Dunwoody and Griffin (1993: 47) point out, this event orientation limits journalistic frames in two ways: it allows news sources to control the establishment of story frames, and it absolves journalists from attending to the bigger environmental picture. Three major types of environmental events can be identified: milestones (Earth Day, the Rio Summit); catastrophes (oil spills, nuclear accidents, toxic fires); and legal/administrative happenings (parliamentary hearings, trials, release of environmental White Papers).

The twin lures of celebrity and symbolism at milestone events can be seen at the 1992 Earth Summit at Rio de Janeiro in Brazil. Those attending included not only more than a hundred heads of state, including US President George Bush, British Prime Minister John Major, German Chancellor Helmut Kohl and Cuban President Fidel Castro, but also an estimated 12,000 representatives from NGOs. Among the celebrities from the world of politics and entertainment were former California governor Jerry Brown, actors Jeremy Irons and Jane Fonda, and American media mogul Ted Turner. Even before the official summit began, a fundamental conflict arose between the wealthy nations of the North and the poorer countries of the South over a wide spectrum of issues. Finally, the summit was accompanied by an array of what *Time* magazine called 'sideshows galore' (Dorfman 1992): a fantasy ballet, *Forest of the Amazon*; an indigenous people's conference; and a concert for the Life of Planet Earth. The symbolism of the occasion was typified by the giant Tree of Life in Rio on which were hung leaf postcards from children worldwide.

Thirteen years later, the 2005 summit of G8 leaders in Gleneagles, Scotland, was more or less appropriated by rock stars Bob Geldof and Bono and converted into a media opportunity to publicise their campaign to eradicate African debt and end global poverty. As a lead-up, ten 'Live 8' concerts were staged across four continents watched by two billion people worldwide. At the meeting itself, Geldof and Bono were accorded quasi-diplomatic status, meeting one-on-one with world leaders, even as the usual assortment of protesters, including environmentalists, were held back behind the barricades. By the time media interest shifted dramatically to the terrorist bombings in London, African poverty had received an unprecedented week in the media limelight.

Catastrophes are the bread and butter of environmental news coverage. They frequently involve injury and loss of life or the possibility of such. There are sometimes acts of tremendous courage or self-sacrifice. Human interest stories abound: the stubborn but proud homeowner who sits on the roof and refuses to evacuate as the floodwaters rise; the baby who is found alive after three days in the rubble of an earthquake-devastated neighbourhood.

According to Wilkins and Patterson (1990: 19), this event-centred reporting is characteristic not only of quick onset disasters such as tornadoes, hurricanes and blizzards but also of slow onset environmental hazards: ozone depletion, acid rain, and so on. In order to fit these latter phenomena into the news agenda, journalists are required to picture them as the recent outcome of an event rather than the inevitable outcome of a series of political and societal decisions.

While event-centred coverage has the advantage of raising public awareness of otherwise ignored environmental topics, it also has a negative side. By focusing on discrete events rather than on the contexts in which they occur, the media tend to give news consumers the impression that individuals or errant corporations rather than institutional politics and social developments are responsible for these events (Smith 1992; Wilkins and Patterson 1990). This is especially applicable for environmental catastrophes. For example, in the case of the 1989 *Exxon Valdez* oil spill, the media framed the story in terms of Captain Joseph Hazelwood's alleged alcohol problems rather than dealing with other potentially important news angles such as the recent history of cutbacks in maritime safety standards administered by the coastguard, or the oil industry's lack of capability in cleaning up large oil spills in settings such as Prince William Sound (Smith 1992). Cottle (1993: 122) has described this as the tendency of an item to remain 'entrapped within the narrow confines of its news format', unable to allow any background explanation or any input from outside, non-official voices.

Furthermore, stories about hazards favour monocausal frames rather than frames involving long and complex causal networks. Thus Spencer and Triche (1994) found that increases in toxic pollution in the drinking-water supply of New Orleans during the summer of 1988 were almost exclusively attributed to a simple natural phenomenon – a drop in water levels in the Mississippi River due to drought conditions – rather than to a combination of low water levels and a long-standing problem with discharges from chemical plants upriver from the city. They speculate that this monocausal framing occurred because newspaper personnel were reluctant to implicate several powerful institutional actors – the US Army Corps of Engineers, the state bureaucracy, the chemical industry – as contributors to this hazard event.

Cottle's comments further suggest a second feature of the news process that shapes the nature of coverage: a public access that is largely restricted to official news sources. Since few reporters themselves feel qualified to sort out the often conflicting scientific, technical and political claims involved in an environmental problem, they either avoid substantive issues altogether (Nelkin 1987) or turn to informed sources[3] who can offer a credible and easily summarised précis of what is happening.

While these 'primary definers' are depicted as coming exclusively from a hierarchy of social and political élites, Cottle (1993: 12) argues that this is not necessarily the case for environmental stories. Analysing a sample of British television programmes from 1991 to 1992, he found that various diverse elements (i.e. scientists, diplomats, local officials and politicians, environmental pressure groups, individual citizens) collectively constituted the primary definers.[4]

At the same time, Cottle indicates that this was by no means 'a situation of open and equal access' since environmental news clearly depends on a number of well-organised interests, some from the dominant élite, some from opposing groups.

However, Anderson (1993b) has questioned whether it is possible to deduce patterns of source dependence from content analysis alone. Supplementing content analysis with interviews, she found that ease of access varies over time. For example, during the late 1980s, Greenpeace and Friends of the Earth had good access to the national media in Britain, but they subsequently experienced some difficulty as the threat to the environment gave way to other issues such as the economic recession.

At various points in its recent history, environmental news coverage has also suffered because it does not fit easily into the structure of routine news production. Metropolitan daily newspapers tend to be partially organised according to fixed 'beats' – city hall, industrial (labour) relations, crime, sports, etc. Schoenfeld (1980: 458) cites one reporter as describing the classic environmental story as a 'business-medical-scientific-economic-political-social-pollution story'. This being so, editors and producers often do not know what to do with stories about the environment. It should be noted, however, that this may have a positive aspect, insomuch as individual environmental reporters are sometimes given considerably more leeway than their colleagues working on other journalistic beats because environmental issues are so often difficult for non-specialists to understand (Fletcher and Stahlbrand 1992: 183).

Smaller newspapers and broadcast newsrooms are less likely to use beats, opting instead for a general assignment system (Friedman 1984: 4). This, however, creates another set of difficulties. General assignment reporters, despite their optimism that they can quickly acquire adequate knowledge about subjects in which they have no background or training, are rarely capable of sophisticated reporting[5] such as that demanded by many environmental stories.

Based on his comprehensive analysis of news coverage of three environmental catastrophes – the 1988 Yellowstone Park forest fires, the *Exxon Valdez* oil spill and the Loma Prieta 'World Series' earthquake in 1989 – Conrad Smith, himself a former photographer and film editor, identifies three major difficulties experienced by such general reporters: (1) they did not conceptualise these major catastrophes as anything more than large-scale versions of warehouse fires or train derailments; (2) they did not have the structural freedom to go beyond the obvious stories; (3) they did not know how to find experts and evaluate their relative scientific qualifications (1992: 190).

When environmentalism first took off as a news story in 1969–70, many daily newspapers set up an environmental beat.[6] Reporters were recruited from allied beats – nature, outdoor recreation, science – or from the general assignment pool. While the volume of environmental coverage rose, the quality did not always keep pace. In particular, these rookie environmental reporters seemed to experience difficulty with both the substance and style of environmentalism (Schoenfeld *et al.* 1979). When the environment faded as an issue after 1970, many of these beats were shut down (Friedman 1983), although some of them were later recommissioned (Hansen 1991).

A final short-term constraint on environmental reporting is the role and influence of news editors. With one eye always fixed on circulation or audience figures, editors tend to favour stories that feature controversy and conflict. As a result, thoughtfulness often gives way to sensationalism. In addition, editors are more likely to be sensitive to external pressures from corporate advertisers and other powerful supporters of the status quo. Reporters know this, and on occasion modify or deliberately overlook significant stories that involve environmental wrongdoing (Friedman 1983). This evidently occurred in the late 1970s in Houston, Texas, where local newspaper reporters were not willing to go against the predominant 'boomtown' mentality and report the problems surrounding a nuclear power plant and a nuclear treatment facility (Hochberg 1980).

Longer-term constraints on environmental journalism relate to historically evolved journalistic priorities, notably the requirements for news 'balance' and 'objectivity'. These dual pillars of objective journalism first arose during the nineteenth century as part of the sweeping intellectual movement towards scientific detachment and the culture-wide separation of fact from value (Gitlin 1980: 268). Despite periodic lapses, newsworkers today still view objectivity and balance as the cornerstones of their profession.[7]

For environmental reporting, objectivity and balance mean that reporters often attempt to distance themselves and their readers from the environmentalist struggle to effect a shift in public consciousness, taking refuge instead in the objectivism of science (Killingsworth and Palmer 1992: 149). Journalists thus see themselves as a neutral and ironic voice, willing to be won over only if the scientific evidence concerning acid rain, global warming, biotechnology, etc. is sufficiently powerful and unambiguous. The major shortcoming of this approach is that few environmental reporters are sufficiently well informed to be able effectively to evaluate the 'scientific standing' (Friedman 1983: 25) of the evidence. Alternatively, reporters may turn to the traditional 'equal time' technique whereby both environmentalist claims-makers and their opponents are quoted with no attempt to resolve who is correct. In this case it becomes difficult for environmentalists to convince the public that an emerging 'issue' is in fact a 'problem'.

Boyne (2003: 35) has identified a tension between the media's dual imperative of analysing risk and creating an appetite for its images. All too often, it is the latter that predominates. Journalists, editors and producers abandon the sceptical stance described above and embrace the role of a 'campaigner'. In such instances, the media actually come to *lead* the public agenda. On occasion, this can lead to considerable harm, especially where the scientific evidence is inflated or misconstrued. This is what appears to have happened in the case of cell phone 'scares' in Britain in the 1990s.

The ideal of objectivity also means that journalists rarely express the content of environmental stories in overtly political terms, opting instead for news frames that emphasise conservation, civic responsibility and consumerism. Lowe and Morrison (1984: 80) even go so far as to contend that a major attraction of environmental issues for the media is that they can be depicted in non-partisan terms, allowing journalists to subversively foster environmental protest at the same time as appearing to maintain a politically balanced stance.

Cottle (1993: 128) echoes Habermas in noting how the media debase the public sphere, refracting the environment through a journalistic prism that reduces politically charged stories such as global warming to the more immediate and mundane domestic and leisure concerns of ordinary consumers; for example, whether a beach holiday is likely this summer.

Constructing 'winning' environmental accounts in the media

As Stallings (1990: 88) has noted, some media accounts of environmental problems drop by the wayside while other 'winning accounts' persist and ultimately succeed in gaining acceptance. Indeed, the media contribute to this by fostering an image of either growth or decay for a particular problem (Downs 1972).

This notion of 'attention cycles' has been examined by McComas and Shanahan (1999). In their content analysis, the researchers analysed stories in the *New York Times* and *Washington Post* from 1980 to 1995 concerning global warming. McComas and Shanahan found that narratives about this environmental issue passed through five stages: a pre-problem stage, a period of alarmed discovery, public realisation of significant progress, gradual decline of intense public interest and a post-problem phase. Narratives about the implied danger and consequences of global warming were more prominent on the upswing of newspaper attention, whereas those dwelling on controversy among scientists received greater attention in the later stages (Dispensa and Brulle 2003: 93). Claims-makers thus need to learn how to keep environmental stories fresh and compelling.

In charting the ascent and tenure of environmental problems on the media agenda, it is possible to identify five key factors.

First, in order to gain prominence, a potential problem must be cast in terms which 'resonate' with existing and widely held cultural concepts (Kunst and Witlox 1993: 4). This is why the frame alignment process is so crucial. Despite over three decades of exposure to environmental discourse, the actual awareness and salience of most environmental issues remains 'pitifully low' (Cantrill 1992: 37). In particular, most citizens, especially in North America, continue to place their faith in science and technology and to believe that economic growth is generally desirable. Thus, packaging an issue in the form of direct criticism of the Dominant Social Paradigm would not appear to be an effective communication strategy for environmental claims-makers. Instead, it makes more sense to situate environmental messages in frames that have wider recognition and support in the target population: health and safety, bureaucratic bungling, good citizenship, and so on.

Second, a potential environmental problem must be articulated through the agendas of established 'authority fora' (Hansen 1991: 451), notably politics and science. If it does not receive this legitimation, a problem will likely stagnate outside the media arena. This was the case in Britain where various 'green' issues (acid rain, ozone damage) lay relatively fallow until invigorated by a speech from Prime Minister Margaret Thatcher to the Royal Society in September 1988, in which she adopted an environmental rhetoric for the first time. The Thatcher speech conferred a new degree of political legitimacy on the environment and the environmental movement and this subsequently diffused throughout many other arenas with the assistance of the mass media (Cracknell 1993).

Third, environmental problems that conform to the model of a publicly staged 'social drama' are more likely to engage the attention of the media than those that do not. As Palmlund (1992: 199) has suggested, the societal evaluation of risk takes the form of a dramatic contest coloured by emotions and containing both blaming games and games of celebration. In this contest there are readily identifiable heroes, villains, victims and even a chorus. Love Canal was the perfect media story by this yardstick with the timid housewife turned activist Lois Gibbs as the heroine, neighbourhood children with their increasing health problems as the primary victims, and Hooker Chemical as the odious polluter.

Some environmental organisations, notably Greenpeace, have been very successful in staging morality plays in front of the global media with themselves as intrepid idealists and a changing cast of characters – whalers, seal hunters, French sailors, nuclear operators – as the villains. By contrast, problems that lack this fairy-tale quality, for example the seepage of indoor radon gas into Canadian and American homes, are more difficult (although not impossible) to sell to the media.

Fourth, an environmental problem must be able to be related to the present rather than the distant future in order to capture media attention. Dianne Dumanoski, an environmental reporter for the *Boston Globe*, notes that some of the more immediate environmental problems such as oil spills interest editors more 'because they can understand that...There's dirty stuff on the rocks; it's not computer models and these guys at MIT talking about something in the future' (Stocking and Leonard 1990: 41).

Global warming appeared to be a far away problem until the abnormally hot summer of 1988 when a series of tangible environmental disasters – droughts, floods, forest fires, polluted beaches – dominated the news. These contributed significantly to *Time* magazine's editorial decision to feature the endangered earth in its Planet of the Year issue of 2 January 1989 (McManus 1989).

Finally, an environmental problem should have an 'action agenda' attached to it either at international level (global conventions, treaties, programmes) or local community level (tree-planting, recycling). Environmental conditions that are less amenable to action are not as likely to appeal to reporters and editors unless, as was the case with the Ethiopian famine, a moral panic can be created around the consequences provoking a flurry of humanitarian relief efforts. Furthermore, rather than advocating some long-term action plan with results which may not be noticed for decades, environmental claims-makers should be able to offer the media some tangible results in the here and now: for example, shutting down an incinerator, cleaning up a polluted harbour, rescuing a beached whale. Unfortunately, as Solesbury (1976: 395) has noted, complex environmental problems with multiple dimensions are the most difficult to process because they can easily become bogged down in scientific disputes and interdepartmental rivalries. In such cases the media will tire of a problem, relegating it to a journalistic limbo where it is considered neither finally retired nor sufficiently topical to be of current public interest.

Mass-mediated environmental discourse

From a topic with no distinct identity of its own, the environment has progressed to the point where it is now an established part of everyday journalism. While there has been a broad upsurge in coverage, there is no single overarching environmental discourse. Instead, the media are the site of multiple outlooks and approaches, some of which are in direct conflict with the others (see Brulle 2000).

At one level, environmental communication is primarily an objectivist scientific discourse. As noted earlier in this chapter, journalists normally view themselves as impartial judges open to conversion only if the scientific proof is seen to be convincing. Scientific claims are reported at face value with relatively little attention to their constructed nature nor to their unknowns and uncertainties (Stocking and Holstein 1993: 202). Journalists have little patience with the thrusts and parries of scientific debate: either a danger exists or it does not.

At the same time, the media routinely lapse into a human interest discourse which 'carries the journalist out of the field of natural science and into the action oriented fields of social movements and politics' (Killingsworth and Palmer 1992: 135). Here, the burden of proof is less exacting. The essence of an environmental problem is more likely to be presented in a single dramatic image: a drum of toxic material, a discarded syringe on the beach, a head of foam on the surface of a trickling stream. Scientific scepticism is replaced by 'common sense'. The emphasis is less on the nature of the conditions that underlie the problem and more on the imputed consequences for people's lives. The narrative is more dramatic, even mythological.

Take for example a wire service story from the mid-1990s (Lawson 1994) on public hearings into a request by a joint Canadian–American venture to convert an unused oil pipeline running through rural Ontario (Canada) to a natural gas conveyance. Rather than examine the technical, economic and environmental feasibility of the project, the reporter chose to emphasise the participation as an intervenor of Jean Lewington, the widow of an area farmer who had spent thirty years successfully fighting a previous pipeline extension, thereby changing the way utility companies must deal with farmers and their land. This was accented by a photograph of Mrs Lewington standing in front of her barn, and a headline that read 'Farm widow re-fights old pipeline foe'.[8]

Third, the media, especially the business press, have increasingly adopted a discourse that presents the environment as an economic opportunity. The key message here is that environmental adversity can be turned into profit through human ingenuity and industry. Much of this type of coverage is product-oriented, touting a wide variety of 'green' products from the energy-saving house to nuts harvested by indigenous peoples in the rainforests of Brazil. The predominant message is that the entrepreneurial spirit need not be incompatible with ecological values; rather, the two are mutually reinforcing. This optimistic view of the environment has been amplified in the rapidly expanding body of stories on the promise and prospects for 'sustainable development'.

Fourth, the media situate the environment as the locus for rancorous conflict. While this environment as conflict package sometimes deals with the wider clash of cosmologies between environmentalists and their opponents, it is more likely to depict these disputes in the same manner as journalists routinely portray industrial relations disputes. That is to say, protesters are implicitly blamed for the disruption of normal commerce, the rationale for their actions is compressed into short sound bites and the background to the conflict is downplayed. The leaders of environmental protest actions are often presented as 'hippies and violent "ecoteurs"' armed and ready for monkey wrenching[9] (Capuzza 1992: 12).

An environmental conflict story may shoot to the top of the news agenda if a well-known celebrity arrives on the scene. For example, the protest against the clear-cutting of the old forest on Clayoquot Sound on Vancouver Island was elevated in news value when Robert Kennedy Junior arrived to 'inspect the damage'. Rancorous environmental conflicts are supercharged with symbolic content with both protesters and their opponents likely to use the framing and reasoning devices identified by Gamson and Modigliani (1989).

One consequence is the spillover of this media discourse into real life ideological battles between environmentalists and their opponents. Thus Dunk (1994) observed that the forest workers in north-western Ontario tend to regard environmentalists as outsiders from 'down south' or from 'big cities', in large part because they uncritically accept the dominant normative structure of the popular media's representation of environmental issues as a confrontation between middle-class, urban-based environmental radicals and local citizens fighting to keep their jobs.

Fifth, the media situate the environment within an apocalyptic narrative. Employing a series of medical metaphors, our planet is depicted as facing a debilitating, perhaps terminal, illness. Overpopulation, loss of biodiversity, rainforest destruction, ozone depletion and global warming are all linked causally to this impending ecological crisis. Despite the caution expressed in scientific media discourse, journalists give considerable news space to the popularised accounts of global threats formulated by Paul Ehrlich (overpopulation,

'Swampy'

One notable exception to the usual negative media coverage accorded environmental dissent is the treatment of youthful protesters against a road-building scheme on the A30 motorway (highway) in Devon, England, in 1997. In fact, a most unlikely hero known only as 'Swampy' arose from this protest. Swampy was the last of five protesters to emerge after camping for a week in a maze of tunnels underneath the road. Among other things, Swampy wrote a column in the *Sunday Mirror* for nine weeks; appeared on a popular TV news quiz comedy show; and was the inspiration for a character in the long-running television soap opera *Coronation Street*. As Paterson (2000: 151) notes, Swampy 'became a byword for environmental direct action and youth disaffection from formal politics'. Paterson argues that the media de-activated the more radical elements in the campaign by *normalising* Swampy and his fellow protesters ('Muppet Dave', 'Animal Magic'). For example, the *Daily Express* dressed Swampy in designer Armani suits for a photo shoot; and the *Daily Mail* profiled Animal Magic as a talented and articulate 16-year-old adolescent who even blushed when asked about her boyfriends. While this may have had the effect of making opposition to the road-building programme seem acceptably idealistic and legitimate (as against coverage of previous protest actions of this type elsewhere in the UK that were depicted as violent and extreme), at the same time it erased 'the connections between road building and broader social and political questions and thus deep opposition of the road protesters to modern forms of organization and power' (Paterson 2000: 158).

biodiversity loss), Steven Schneider (global warming) and Norman Myers (tropical deforestation) and other prophets of doom. Thus *Time* magazine subtitled its 1989 special issue cover story on the greenhouse effect with the caption, 'Greenhouse gases could create a climatic calamity' (Killingsworth and Palmer 1992: 158).

Finally, the environment is scrutinised through the lens of institutional decision-making. Rather than attributing it a unique status, the environment is treated as another policy area alongside health care, education and social services. The focus here is on regulatory agencies and processes, impending legislation, political personalities (Al Gore, Maurice Strong) and international fora (United Nations, European Union). Too often this leads to an ingrown policy debate between political and scientific élites (Wilkins and Patterson 1990: 21) in which the public is only an incidental bystander.

At any one time, various media packages as well as a plethora of individual news frames may compete for dominance. A single environmental event may have multiple shifting frames as it develops. For example, Daley and O'Neill (1991) trace the odyssey of the *Exxon Valdez* oil spill from a disaster narrative (the public as helpless victims, a catastrophe outside human control) through a crime narrative (the captain was culpable) to an environmental narrative (environmentalists contested the statements and practices of industry and government officials). At the same time, attempts to frame a story may fail. In the *Exxon Valdez* case a competing subsistence narrative (the oil spill posed a threat to native Alaskans' way of life) was all but ignored, appearing only in an indigenous publication, the *Tundra Times*. Journalists are thus faced with choosing from an assortment of narratives, languages and viewpoints at the same time as adhering to the formats and structures imposed by standard journalistic practice.

Conclusion

What should be evident from the discussion in this chapter is the considerable extent to which environmental news is socially constructed. In large measure this is a reflection of the rhythms and constraints inherent in the practice of journalism itself. In addition, it reflects the multiple competing claims that newsworkers must routinely sort out in the course of putting together a story. This central difficulty in reporting has ben summed up by Stocking and Leonard in this way:

> The environmental story is one of the most complicated and pressing stories of our time. It involves abstract and probabilistic science, labyrinthine laws, grandstanding politicians, speculative economics and the complex interplay of individuals and societies. Most agree it concerns the very future of life as we know it on the planet. Perhaps more than most stories it needs careful, longer-than-bite-sized reporting and analysis now.
>
> (1990: 42)

Whether this depth of coverage is realistically possible is an open question which depends on several factors.

First, editors and producers, the newsworkers (and gatekeepers) who effectively set everyday line-ups and assignments must see environmentalism as more than a transient phenomenon which loses its lustre once it ceases to register strongly in public opinion polls and government agendas. This is less likely to be the case in regions of the country where environmental conflict is endemic because of a natural-resource-based economy. Ironically, the one section of the media where environmental coverage has become institutionalised is in the financial pages where 'green business' is seen as having increased relevance.

Second, environmental issues must be perceived as occupying a distinctive story niche rather than simply overlapping a multitude of existing subject areas – politics, business, agriculture, science and technology. Without a distinctive image, environmental coverage is destined always to remain event-driven and conflict-oriented. At the same time, environmental problems are by their very nature intricately tied in to economic and political structures and policies, making it difficult and sometimes even inadvisable to consider them separately; for example, this is the case with many Third World 'sustainable development' stories. It is thus difficult to balance the need for a distinct environmental specialty beat with the need for a depth of coverage that may reside in other areas of journalistic expertise.

Finally, some way must be found to combine 'muck-raking' or 'exposure journalism' with the longer-term goals of environmental education and policy reform. Investigative reports in the press or on television programmes such as *60 Minutes*, *Frontline* or *The Fifth Estate* may temporarily shock audiences but they do not necessarily result in either a deeper understanding of an issue nor in effective regulatory action. Indeed, sometimes there can be a response quite different from that desired by activist claims-makers. Fletcher and Stahlbrand (1992: 195) cite what occurred in the early twentieth century when Upton Sinclair wrote a widely noticed exposure of the exploitation of immigrant workers in the large meat packing plants of Chicago in his book *The Jungle* (1905):

> His dramatic example of a man falling into a machine and being minced with the meat led not to a better protection for workers but rather to meat inspection laws, a reform the meat packers wanted to help them compete in European export markets.

In a similar fashion, a segment on *60 Minutes* concerning a community activist's fight against an incinerator which, she charged, was emitting toxic pollutants evidently resulted in a number of positive business enquiries from other American municipal governments to the waste management company which operated the facility. There must, then, be some blend of story elements that succeeds in raising an alarm in the public arena and then situating this concern within a clearly defined set of goals for environmental reform.

7 Science, scientists and environmental problems

It is rare indeed to find an environmental problem that does not have its origins in a body of scientific research. Acid rain, loss of biodiversity, global warming, ozone depletion, desertification and dioxin poisoning are all examples of problems which first began with a set of scientific observations. Ultimately, it is the scientific underpinnings of these environmental problems that lift them above most other social problems that are more dependent on morally based claims (Yearley 1992: 117).

Furthermore, scientific researchers act as 'gatekeepers', screening potential claims for credibility. In 1988, when the British organisation, Ark, mounted a publicity campaign in which they alleged that melting ice-caps due to global warming would raise sea levels five metres in sixty years, thereby covering much of Britain with water, more sober scientific estimates of less than a metre rise quickly discredited the Ark initiative (Pearce 1991: 288–9).

Yet paradoxically, science itself is frequently the target of environmental claims. One notable example of this is the contemporary debate over genetic engineering and its potentially harmful effects in the environment. In cases such as this, claims-makers explicitly reject the technical rationality of science in favour of an alternative cultural rationality that appeals to 'folk wisdom, peer groups and traditions' (Krimsky and Plough 1988: 107). Science is pilloried for interfering with the natural order rather than lauded for lending its authority to a claim.

Science as a claims-making activity

The profile of science presented so far would seem to suggest that scientific findings reflect the physical reality of the natural world in a relatively straightforward manner. Science would therefore appear to be a search for truth in which the goal is to obtain a clear reflection of nature, as free as possible from any social and subjective influences that might distort the 'facts'.

Yet to the contrary, the assembly of scientific knowledge is highly dependent on a process of claims-making. In this regard, Aronson (1984) identified two types of knowledge claims made by scientists: cognitive claims and interpretive claims.

Cognitive claims aim to convert experimental observations, hypotheses and theories into publicly accredited factual knowledge. Blakeslee (1994) describes this conversion process as one in which scientists must adeptly stake novel claims while at the same time fitting them into an established research tradition. She gives as an example the process of cognitive claims-making in the physics journal, *Physical Review Letters*, in which contributors' letters announcing innovations have come to resemble journalistic accounts of scientific findings complete with an arsenal of rhetorical strategies.

Interpretive claims, on the other hand, are designed to establish the broader implications of the research findings for a non-specialist audience. Interpretive claims implicitly ask lay audiences to certify the social utility of the research, and the content of the claim supplies the reason they should do so. For example, in the case of global warming, the cognitive claim is that gases from cars, power plants and factories are creating a greenhouse effect that will boost the temperature significantly over the next seventy-five years or so. The interpretive claim here is that this heating trend is potentially dangerous because, among other things, it will cause havoc with the existing geography of the Earth, flooding some low-lying areas such as the Netherlands and New Orleans and bringing drought to fertile agricultural regions such as the American Midwest. In the wake of the tremendous destruction wreaked by Hurricane Katrina in 2005 on New Orleans and the Gulf Coast, this predictive claim took on an enhanced credibility and resonance.

Not only do scientists make knowledge claims but they also routinely construct 'ignorance claims' (Smithson 1989). This means that researchers highlight 'gaps' in available scientific knowledge in order to make a case for further research funding or, conversely, to retard further policy action on the grounds that not enough hard data exist to justify regulation or legislative activity (Stocking and Holstein 1993).

Aronson (1984) outlines three types of interpretive claims which scientists make: technical, cultural and social problem.

Technical interpretive claims-making occurs when researchers act as scientific advisers to industry and government. This often involves the evaluation of risks posed by controversial technologies (nuclear power, genetic engineering), suspected toxic pollutants (dioxin, mercury) and global hazards (ozone depletion, global warming). While in theory scientific advisers are restricted to a narrow technical assessment role, in reality they incorporate their own political agendas and knowledge claims into their own interpretations and recommendations.

Salter (1988) uses the term 'mandated science' to refer to the science which is used for the purposes of formulating public policy including studies commissioned by government officials and regulators to aid in their decision-making. Despite an official face of neutrality flowing from scientific expertise, members of expert panels regularly make moral and political claims and choices. These choices are fashioned as much by policy considerations as by scientific norms. For example, a scientific advisory committee dealing with pesticide safety may be equally aware that banning a chemical compound will negatively affect a $500 million industry, while recommending its use could have serious health effects that will only become evident ten years later. This knowledge, Salter observes, affects the committee's recommendations as much as does their technical data, thereby imbuing their activities with a strong interpretive flavour.

Cultural interpretive claims attempt to develop ideological support both for expenditures on scientific research and for the autonomy of science. The media through which the claims are presented are public speeches, articles in popular scientific magazines (*New Scientist*, *Scientific American*) and on the opposite-editorial pages of influential newspapers (*New York Times*, *Washington Post*, *The Times* (London)), testimony before parliamentary enquiries and participation in government–industry committees and panels. In some cases, the receipt of an international scientific prize allows the researcher a unique platform from which to address broader social and political concerns. This is what occurred in Canada when John Polanyi won a Nobel Prize in chemistry and took advantage of the outpouring of public attention to address a raft of issues from government underfunding of universities to nuclear disarmament and peace. In other cases, the threat of a public review of scientific work can mobilise scientists towards making cultural interpretive claims. For example, Krimsky (1979) has demonstrated that the threat of external intervention and control into recombinant DNA molecule research in the 1970s turned American scientists into surprisingly effective lobbyists for scientific autonomy and the freedom of self-regulation.

Social problem interpretive claims assert the existence of a social problem that a particular scientific speciality is uniquely equipped to solve. Aronson identified three conditions under which scientists are likely to make such claims.

The first is when a new discipline has no foothold in the academic world and therefore must appeal to external constituencies to obtain funding and political support for its work. To a degree, this has been the case for environmental science, which has been routinely criticised by many mainstream scientists for doing research that is defensive or of low quality (Rycroft 1991).

The second condition is when enterprising scientists, ever in search of new publicly derived research funds, attempt to show that their existing research work contributes to the solution of a recognised social problem or that it will successfully solve a previously unrecognised problem. This was characteristic of cancer research in the 1970s and AIDS research in the 1980s.

Figure 2 Global warming fact or fiction debate

Source: *Saturday Evening Post*, 264(5), September/October 1992

A third condition under which social problems claims-making is likely to occur is when scientists are confronted by social movements which seek to restrict their research. In this situation, scientists are compelled to assemble and promote their own set of interpretive claims to justify either why a problem exists and their research should continue or why their research should not be construed as constituting a problem.

Aronson argues that there is a tendency for the first two forms of interpretive claims, technical and cultural, eventually to be transformed or subsumed by the social problem form because what is basically at stake is the social utility of science. That is, researchers recognise that it is better strategy proactively to make a case for the social benefits of their work rather than to wait and subsequently have to justify it in an atmosphere of scepticism and budget slashing.

Scientific uncertainty and the construction of environmental problems

What particularly opens the door to the creation and contestation of environmental problems is the inability of science to give absolute proof – unequivocal evidence of safety. Instead, scientists are reduced to offering estimates of probability that often vary widely from one to another. This lack of certainty allows claims-makers both within and outside science to assert that the situation is alarming, that the risk is too high and that society should do something about it.

Furthermore, mainstream science and green activists differ fundamentally as to when human intervention is necessary to protect the environment. This difference in perspective is nicely illustrated in a debate that took place in the early 1990s in the pages of the British science magazine, *New Scientist*.

Brian Wynne and Sue Mayer argue that the decision whether to take official action on environmental risks should be governed by a *precautionary principle*. This states that if there is reason to suspect that a particular substance or practice is endangering the environment then action should be taken even if the evidence is not ironclad. The rationale behind this view is that it will be too late to respond effectively if we wait for a final scientific resolution years down the road. Where the environment is at risk, there is, they argue, 'no clear cut boundary between science and policy' (Wynne and Mayer 1993: 33).

The precautionary principle has been enormously influential, especially in Europe. British sociologist Adam Burgess (2003: 105), who views the concept as problematic, nonetheless acknowledges that it 'forms the basis for much domestic and international policy making' and, in its harder form, 'represents a frontal challenge to the experimental method that has been so central not only to science, but to modern society in general'. Theofanis Christoforou (2003: 205–6), a legal adviser to the European Commission, observes that in the EU, the precautionary principle has the status of nothing less than 'a mandatory treaty principle'. If properly applied, he says, it 'can be deployed to ensure that the societal values and democratic policy choices on health and environmental protection are fulfilled'.

The opposing position is presented by Alex Milne, a consulting chemist who spent 34 years working in the paint industry. Milne rejects the precautionary principle, which he labels as one of the central doctrines of 'green science', as entirely the wrong approach. It is worse, he claims, than the legal principle in *Alice in Wonderland*, where the pattern was 'sentence first, verdict afterward'; here it is 'verdict first, trial afterward and no need for evidence' (1993: 37). The precautionary principle, he complains, has nothing to do with science: it is entirely an administrative and political matter.

A large measure of the disagreement here revolves around how science should be done. In traditional science, a reductionist principle predominates. This means that researchers break down a problem into the smallest number of constituent parts and look at each part separately, controlling as much variation as possible. If you want to look at the effect of a toxic chemical on the breeding pattern of fish, you isolate the fish in an experimental setting, vary the levels of the chemical and record the birth results. By contrast, a cardinal principle of green science is the necessity of looking at the world holistically. Since everything is connected to everything else, it does not make sense to disassemble an ecological web experimentally. For example, immunity is a complex system that is linked to a variety of factors from genetics to environmental pollution to socio-psychological stress. Causation may be indirect or multiple, making it all but invisible to the reductionist perspective of traditional 'good science' (Wynne and Mayer 1993: 34).

In policy terms, good science manifests itself in the form of an *assimilative approach* which purports to define scientifically the capacity of an ecosystem to assimilate pollutants without harm and then licensing industrial discharges within these 'proven' safe limits. What this ignores, environmentalists maintain, is the possibility of a chemical interaction among the polluting chemicals that creates a potential for end effects not anticipated by the assimilative model.

As Salter (1988) has observed, quite different sets of criteria are applied depending on the context in which research evidence is evaluated. Conventional science possesses a deeply ingrained capacity to handle ambiguity; indeed, most journal articles routinely end with the caveat 'further research is needed'. By contrast, the burden of proof is stricter when scientists appear before regulatory hearings or in the courtroom. Here, legal concepts such as 'reasonable doubt' are prominent – anathema to scientists who are socialised to always couch their conclusions in conditional terms. In this regard, Yearley (1992: 142) points out that scientific expertise depends on elements of judgements and craft skill, informal aspects of science which can be highlighted in a legal or regulatory hearing to make scientific evidence appear like mere opinion. This tendency is even further exaggerated when environmental groups communicate using a moral discourse in a setting where the conventions of a scientific, legal or regulatory discourse predominate. The precautionary principle is a good example of an environmental principle that operates on a different plane of certainty than do societal control institutions.[1]

The crucial dilemma, then, is that social problem interpretive claims which rest on sound scientific evidence are generally more 'robust' than those claims only supported by opinion (Yearley 1992: 76) but there is a fundamental disagreement between environmentalists, scientists, regulators and legalists over what constitutes sound scientific evidence.

Blowers (1993) has observed that scientific evidence is problematic as a basis for environmental policy-making in five ways. First, there is the problem of cause and effect that we have been discussing; this makes it difficult to establish responsibility for the externalities produced by polluting activities. Second, there is the problem of forecasting impacts; for example, the uncertainty about the incidence, distribution, timing and effects of global warming. Third, uncertainty over the consequences of present actions and the risks imposed on future generations may lead to a paralysis of policy or to a tendency to discount the future risks of present action. Sometimes, in fact, another future focused scenario – the crushing burden of a spiralling national debt – may discourage taking bold ameliorative or prophylactic steps in the here and now. Fourth, the frequent absence or sparseness of environmental data not only makes it more difficult to provide sound scientific judgments but it also opens the door to manipulation by vested interests who claim that environmentalists have exaggerated the danger. Finally, the fragile interpretations of environmental science can easily run aground on the shoals of politics where conflicts between interests dominate. This is especially the case where one is dealing with broad speculative ideas such as the Gaia hypothesis rather than narrower, more empirically captured linkages.

Identifying environmental problems as scientific issues

Seldom does an environmental problem pop up overnight with no past legacy of scientific observation and debate. Rather than grow along a linear path, the process by which environmental problems are identified and evolve as scientific issues is characterised by the creation of a pool of knowledge that expands serendipitously in unexpected directions (Kowalok 1993). Individual pieces of data in this pool may be generated through projects that employ the reductionist methods of traditional science, but in the end it is a flash of holistic insight that leads to final understanding.

Despite appearances to the contrary, the basic outline of many environmental problems has been around for a long time. For example, the theory that greenhouse warming is caused by human generated emissions of carbon dioxide has been known for more than a century but the greenhouse effect was not considered a priority problem until the 1980s (Cline 1992: 13–14). Similarly the term 'acid rain' together with many of its fundamental principles was first introduced by chemist Robert Angus Smith in 1872 but did not emerge as a full-blown scientific problem until the 1970s.

What then propels an environmental problem of long standing into a current scientific claim of critical proportions?

First, the real or perceived magnitude of the condition may suddenly rise to 'crisis' proportions. For example, species extinctions have been increasing steadily since 1600 as human settlements have spread across the globe. Recently, however, it has been claimed that we have seriously tipped the balance between the appearance of new species and the extinction of existing ones (Tolba and El-Kholy 1992). At the same time, the loss of old growth forests and plant and animal species captures the attention and concern of conservation biologists and other scientific claims-makers precisely because these natural resources are down to their last twenty, ten or one per cent, making preservation appear more crucial.

Second, new methodologies, research instruments or data banks may allow scientists to come to conclusions that were impossible earlier on. For example, data provided by the European Air Chemistry Network starting in the 1950s allowed Swedish researcher Svante Oden to advance his pioneering theories about acid rain, while James Lovelock's comparisons of the concentrations of fluorocarbons in the lower atmosphere with annual amounts of industrial production opened the door to chemists Mario Molina and Sherwood Rowland to document the key link between CFC products and ozone destruction (Kowalok 1993).

Third, the holistic character of global ecosystems means that rising scientific and public interest in one environmental problem readily generates interest in another interrelated problem. Thus scientific concern over tropical deforestation has spread well beyond the boundaries of silviculture due in large part to the key role which the loss of tropical forests plays in what are presently the two highest profile global environmental problems: global warming and the loss of biological diversity. Mazur and Lee (1993) illustrate this in schematic fashion, demonstrating how the rise of public concern over the problem of the global environment is actually a weaving together of several strands of concern over specific problems, each of which has arisen at a different point in time. This synergy is not, of course, always readily apparent and scientific entrepreneurs may need explicitly to establish the relevance of one issue for another.

Fourth, the establishment of official research programmes, centres and networks may create a hothouse in which research into an environmental problem may be successfully nurtured, even if this is not the original intention. For example, the decision in December 1979 by the Council of the European Community to establish a multiannual research programme in the field of climatology was taken in part because of concern about what was essentially a regional problem – the 1976 drought which affected some African and European areas. Once in place, this programme became both the focus of foundation-building research on the physico-chemical processes related to the increasing concentrations of greenhouse gases in the atmosphere and a source from which scientific findings and terms such as 'greenhouse effect' and 'climate change' circulated outwards into EC policy-making circles (Liberatore 1992).

In all of this the identification and characterisation of threats is highly dependent upon a network of international scientific conferences and collaboration (Kowalok 1993: 36–7). Not only does this permit researchers to learn new methodological techniques or to find the missing pieces in their own puzzles but it also helps build their confidence that they are not alone, an especially important shot of morale boosting when a theory seems radically new and controversial. This was very much the case with the groundbreaking research on the acid rain problem where Canadian and American researchers did not fully appreciate the global relevance of their own findings until they came face to face with similar findings from Scandinavia as presented by Svante Oden on his 1971 lecture tour of North America (Cowling 1982).

Coming out: communicating new environmental problems to the world

The transition from cognitive to interpretive scientific environmental claim is comparable to a 'coming out' in which the ingénue makes a public representation of identity. At some point, the circulation of information around an essentially closed scientific loop is interrupted and the urgency and salience of a problem is shared with the outside world.

One common way of doing this is to convene a public forum at which a mixture of scientists, environmentalists and administrators jointly address the various dimensions of the problem in the full glare of the media spotlight. Alternatively, a claim may be articulated at a congressional or parliamentary hearing where media coverage is usually assured. For example, the 1981 US Congressional testimony by Peter Raven and Edward Wilson was important in establishing the economic utility of preserving endangered species of insects such as the butterfly or the honey bee, particularly for the development of new crops, pharmaceuticals and renewable energy sources (Kellert 1986). Similarly, the ozone depletion issue in Britain was not launched until parliamentary hearings were held in early 1988; strong representations were made in both Houses of Parliament to the effect that the United Kingdom must became a world leader in the drive to protect the ozone layer (Benedick 1991). A third channel for the dissemination of newly constructed scientific environmental problems is a scholarly conference at which reporters from major newspapers are present looking for 'blockbuster' theories. This is what occurred in September 1974, when the *New York Times* picked up on a delivered paper dealing with the threat of CFCs to the ozone layer; the *Times* article 'signaled the beginning of public concern over CFCs and their use in aerosol cans and refrigerators' (Kowalok 1993: 19).

In other cases, however, this process is short-circuited when scientific entrepreneurs go directly to the media. Svante Oden, the Swedish soil scientist who first proclaimed the theory of acid rain, published an account in the Stockholm newspaper *Dagens Nyheter* a year before he published in a scientific journal and five years before the issue arose at the 1972 UN Conference on the Human Environment. Similarly, in Germany, biochemist Bernhard Ulrich's hypothesis that huge tracts of German forests would be dead within five years due to damage from acid rain was presented as established fact in an article in *Der Spiegel*, a mass circulation periodical, provoking widespread national alarm.

How effective one channel is compared to another depends on a number of factors. If there is no consensus among scientists themselves and strident opposition from industry, a more individual approach may work best. Despite periodic attempts to raise the issue, the problem of pesticide poisoning in the United States was being suppressed[2] until Rachel Carson published her indictment in *Silent Spring*. Subsequently, a number of scientists came forward in her defence and the problem was legitimated when, in May 1963, a special panel of the President's Scientific Advisory Committee released a report that was critical of the pesticide industry. On the other hand, jumping the gun before scientific consensus has been established may succeed in capturing media and public attention but at the risk of bringing peer censure by fellow scientists. This occurred in 1988 when James Hansen, director of the NASA Institute for Space Studies, testified before a US Senate committee that summer heatwaves such as that which was being experienced at the time were directly attributable to the greenhouse effect. This norm within science against premature revelation has no doubt been strengthened as a result of the 'hoax' over cold fusion in which the researchers announced their findings at a press conference in Utah prior to subjecting them to peer review.

Science and environmental policy-making

In order for a scientific issue to become policy it must be translated into something that is 'treatable'. As a result, at the policy formulation stage the contribution of natural scientists usually diminishes while the role of socioeconomic and technical experts grows. For example, Liberatore (1992) found that while natural science findings still played an important role in the international debate on global warming in the early 1990s, it was the input of economists, policy analysts and energy technology experts that was crucial in shaping the nature of the European Community response.

This relationship between science and policy-making has best been captured by political scientists by using two concepts: epistemic communities and policy windows.

Epistemic communities

Haas has described the contribution of 'epistemic communities' as critical in achieving international cooperative agreements on environmental issues.

Epistemic communities are 'transnationally organized networks of knowledge based communities'; that is, internationally linked groups of specialists who offer technical advice to political decision-makers.

What gives them a key role in a process usually closed to non-politicians is the uncertain nature of environmental problems. Political leaders may be highly skilled in negotiating trade pacts or treaties but they feel at a distinct disadvantage in dealing with planet-threatening conditions relating to atmospheric shifts or chemical overloads. Under such circumstances information is at a premium as a strategic resource, and politicians, in order to reduce such uncertainty, ' may be expected to look for individuals who are able to provide authoritative advice on whom to pin the blame for a policy failure or as a stop-gap measure to appease public clamour for action' (Haas 1992: 42).

Epistemic communities, Haas contends, are not only bound together by a technical expertise but they also share a number of causal and principled beliefs. In the case of environmental issues, these communities of knowledge were initially composed of ecologists who share a belief in the need for a holistic analysis – a view which carries over to the policy advice that they give. For example, this was characteristic of an epistemic community of ecologists and marine scientists who spearheaded intergovernmental efforts in the 1980s to control pollution in the Mediterranean Sea (Haas 1990).

An epistemic community has the capacity to be influential both in defining the dimensions of a problem and in identifying likely solutions. Thus, Haas demonstrates how a transnational epistemic community of atmospheric scientists was successful in influencing the negotiations that led to the signing of the Montreal Protocol on the protection of the ozone layer in 1987 by 'bounding discussions on the broad array of substances to be covered and the rapidity of regulations' (Haas 1992: 49). Once the epistemic community has laid out the basic parameters of the settlement, it is up to the political leaders to decide what compromises have to be made in order to obtain agreement.

One especially influential conduit through which an epistemic community can shape the policy process is the international scientific assessment panel. Citing the examples of the Millennium Ecosytem Assessment, the Global Biodiversity Assessment and the Intergovernmental Panel on Climate Change (IPCC), Mooney (2003: 49) identfies five features that give these fora high credibility: (1) these panels carefully evaluate peer-reviewed literature; (2) they usually provide some measure of the certainty of the conclusions they draw; (3) the participants are balanced in expertise, region and gender; (4) the results of the assessment undergo rigorous review at many levels, and (5) the final document puts the technical findings into terms that are relevant to policy.

These assessment panels, however, are not necessarily fully representative of all researchers or of the full spectrum of scientific claims pertaining to a particular controversy. For example, the IPCC, whose considerable importance is that it provides the scientific consensus and legitimacy that underpins the Kyoto Protocol, is said by some to disproportionately favour the views of the climate-modelling community found in a handful of large research laboratories often associated with meteorological offices (Boehmer-Christiansen 2003: 81).

Not only do these modellers differ from other researchers working on the global climate change issue, but also they vary internally as well. In his ethnographic study of the *epistemic lifestyles* – the strategies and assumptions they use to build and validate models of the climate – Simon Shackley (2001) demonstrates that there is considerable variation in the modelling styles and therefore in the kinds of knowledge claims associated with differing national and laboratory cultures (Miller and Edwards 2001: 20). These results show that scientific certainty cannot be divorced from epistemic lifestyle, but, instead, must be 'negotiated'. The existence of epistemic lifestyles, Shackley concludes, suggests one important source of diversity in the practice of climate science and indicates that agreement on the role of human-induced global warming may be somewhat less uniform than is often assumed.

Not all political analysts agree with the elevation of Haas's scientific coalitions to a central place in the environmental decision-making process. Haas's model is said to break down in the degree of autonomous power accorded to the epistemic community. That is, scientific coalitions can use their resources to highlight a problem but they must enlist political leaders from their individual nations to have a real impact on treaty negotiations. These leaders may find it advantageous to engage in international problem solving but ultimately they are guided by domestic political considerations (Susskind 1994: 74–5).

Individual governments depend on the technical expertise built up by environmental movement organisations such as Friends of the Earth, Greenpeace and Pollution Probe. In recent years, these groups have devoted considerable resources to building up their own in-house research capabilities, hiring scores of bright, young, idealistic PhDs fresh out of graduate school. In addition, conservation and environmental organisations typically have scientific advisory committees and call upon the voluntary support of university scientists and civil servants who are scientists (Yearley 1992: 126). As a result, there is a synergy between organisations and official policy-makers who find the knowledge and information produced by Greenpeace and others to be of considerable value in staking out their position in public arena debates over environmental issues (Eyerman and Jamison 1989; Lowe and Goyder 1983).

While epistemic communities may be international in scope, the centre of gravity for scientific claims-making on specific issues tends to reside in a specific nation. For example, it was US scientific leadership that propelled the ozone depletion problem into global prominence while Swedish (and Norwegian) research on acid rain was vital in elevating that issue to problem status. In the former case, a critical infrastructure clearly existed as the result of the space programme and the pre-eminence it gave to the United States in researching the stratospheric sciences. This was particularly located in two government agencies – NASA (National Aeronautics and Space Administration) and NOAA (National Oceanic and Atmospheric Administration) – as well as in the graduate faculties of major American universities (California, Harvard, Michigan). When researchers at these institutions voiced concern about events in the stratosphere, the site of the ozone problem, the media and the general public as well as political leaders tended to pay attention (Benedick 1991). In the case of acid rain, the forests and lakes were seen as a vital component of the Swedish economy and therefore were accorded high research priority. When the transnational origins of acid precipitation became obvious in the research data reported by Oden and others, the Swedish government did not hesitate to present these findings aggressively at the 1972 Stockholm Conference.

Policy windows

Another political science model that can be used to link science and policy-making is Kingdon's 'garbage can' model. Adapted from a model of organisational choice developed by James March and his colleagues, this proposes the operation of three major process streams in government agenda setting: (1) problem recognition, (2) the formation and refining of policy proposals, and (3) politics. These three streams usually develop and operate largely independently of one another. However, at critical times the three streams may come together or 'couple'. Kingdon describes this as the opening of a 'policy window' and attributes the main responsibility for this action to 'policy entrepreneurs' within the political system. Individual entrepreneurs do not open the window but they take advantage of the opportunity once it has occurred. At key junctures, then, solutions become joined to problems and both are joined to favourable political forces.

Hart and Victor (1993) have employed Kingdon's model to explore the role of scientific élites in influencing American policy on climate change for the years 1957–74. In their interpretation, science, policy and politics evolve in separate unconnected streams creating both solutions in search of problems and political problems in search of solutions. Scientific élites, assuming the entrepreneurial role, identify policy windows and seize advantage of them.

This is what occurred in the United States in the 1970s. For the better part of twenty years, two interesting scientific discourses relating to the climate had been meandering along, attracting some support but unable really to get moving in terms of either funding or public recognition. These were the 'carbon cycle discourse' which addressed the question of whether and why atmospheric concentrations of carbon dioxide (CO_2) were increasing and the 'atmospheric modeling discourse' which asked what would happen to the climate if higher concentrations of CO_2 were reached. The former discourse was coordinated by an oceanographer, Roger Revelle, while the latter was promoted by John von Neumann, the father of scientific computing.

In the early 1970s, the rise of the American environmental movement created a policy window that these élite scientists successfully exploited in order to mobilise financial and political support and raise public awareness. Hart and Victor (1993: 661) describe this as a synergistic relationship in which scientific findings such as those relating to the greenhouse effect 'catalysed the rebirth of environmentalism' while at the same time environmentalism 'acted as a midwife for new scientific agendas – legitimating them and providing constituencies for their results'. Especially influential in linking the two research streams was Carroll Wilson, an MIT management professor, who was the guiding spirit behind the publication in 1970 of a report, entitled *Study of Critical Environmental Problems*, which was explicitly interdisciplinary and environmentalist in tone.

Hart and Victor (1993: 668) emphasise that very little new scientific information about the prospects of global warming was produced between the late 1960s and the early 1970s.

The 'hockey stick'

In making the case that human activity in the industrial era is primarily responsible for global warming, one very powerful promotional tool has been a graphic nick-named the 'hockey stick'. Like the 'hole' in the ozone layer, this commands public attention by presenting a visual image that is easy to identify and recall. The graph is a reconstruction of temperatures over the past millennium, assembled from data from tree rings, corals and other markers. For most of this period, there are evidently only relatively small fluctuations in temperature (the stick shaft). Then, at the beginning of the twentieth century, there was a sharp upward movement (the blade of the stick).

First published in a 2001 report by the United Nations Intergovernmental Panel on Climate Change (IPCC), the hockey stick graph has been replicated in presentations and brochures used by hundreds of environmentalists, scientists and policy-makers (Regalado 2005: A1). The Canadian government, for example, promoted the hockey stick on its website, sent it to schools across the country and cited it in pamphlets mailed out to all Canadians (McIntyre 2005: FP 19).

Lately, however, the hockey stick graph has come under attack. This critique initially emerged from an unlikely source: Stephen McIntyre, a semi-retired Canadian financial consultant to small minerals-exploration companies, who claims he found some basic mathematical errors in the model. Together with Ross McKitrick, a Canadian economics professor and fellow 'climate sceptic', McIntyre began to publish his views, first in the British social science journal, *Energy & Environment*, whose editor Sonia Boehmer-Christiansen is known for her maverick views on the global warming issue, and then in *Geophysical Research Letters*, a peer-reviewed scientific journal. Climatologist Michael Mann, lead author of the original article in *Nature* where the the hockey stick graph was introduced, has denounced the McIntyre-McKitrick critique of his work as 'frivolous' and politically motivated (Regalado 2005: A13). Nonetheless, Dr Mann and his co-authors were compelled to publish a partial correction in *Nature*, although they still maintain that the data are sound.

While the majority of the international scientific community continues to endorse the IPCC Report, there are pockets of dissent. Some question whether the hockey stick graph significantly underestimates temperature fluctuations prior to the twentieth century, most notably in the years around AD 1000, and again between 1400 and 1600. Others maintain that Dr Mann and his colleagues erred in relying heavily on US bristlecone pine records, misinterpreting a hockey stick configuration as a temperature signal, rather than as evidence of a different biophysical trend. In reply, climate specialists who support Mann's conclusions point to a host of other indicators that the planet is warming up, from receding glaciers in Alaska to Mount Kilimanjaro, stripped of its snow cap for the first time in 11,000 years (Lovell 2005).

Rather, what was different was that the two lines of research were brought together in a new, redefined, scientific agenda that was then successfully sold to political decision-makers and to the news media as a global environmental 'pollution' problem. As we discussed in Chapter 5, this presentation will be enhanced if a simple, visual metaphor is utilised. The 'hole' in the ozone layer is one example of this. Another, more controversial one is the 'hockey stick' graph that has been used to make the case that temperatures are spiking upwards since the beginning of the twentieth century and this is evidence that human activity in the industrial era is causing dangerous global warming.

Scientific roles in environmental problem-solving

Susskind (1994) has proposed five primary 'roles' which are played by scientific advisers in the environmental policy-making process: trend spotters, theory builders, theory testers, science communicators and applied policy analysts. These roles frequently overlap but each has its own tasks and agendas.

Trend spotters are scientists who are the first to detect changes in ecological patterns and to understand their significance correctly. Occasionally, the trend spotter may be a lone scientist who observes some important pattern in the micro-ecology of the pond or marsh and is able to extrapolate this onto the larger environmental canvas. More common, however, are trend spotters who are part of a scientific team that is engaged in gathering and analysing longitudinal data such as that assembled from the LANDSAT satellite or from the European Air Chemistry Network.

Theory builders try to explain the causes for the changes that the trend spotters identify. They are inclined to engage in model building, both to fit explanations to past circumstances and to predict future effects.

Theory testers critically scrutinise the models suggested by theory builders. Using pilot tests or controlled experiments, they attempt to ascertain whether the hypotheses and propositions generated by the model can be empirically proven.

Science communicators attempt to translate difficult-to-decipher data into terms that the public at large can understand. They are key players in the 'coming out' process that was discussed in an earlier section of this chapter. Some communicators such as Edward Wilson are eminent scientists who feel a strong moral responsibility to bring the fruits of their research to the public. Others, for example, the Canadian geneticist and broadcaster David Suzuki, are researchers who have made a conscious decision to spend their life popularising science and carrying the ecological message to a wider audience.

Applied policy analysts act as consultants to political decision-makers, converting scientific findings into policy recommendations. They play a prominent role in the formulation of environmental treaties because they take what is often abstract scientific information and recast it in terms that are amenable to legislation or to international agreements.

Each of the five types of scientists may contribute throughout the environmental problem-solving process but there is a considerable degree of specialisation; that is, trend spotters and theory testers are usually more prominent during the fact-finding stages while science communicators and policy analysts play key roles during the negotiation/bargaining period (Susskind 1994: 77). In terms of the three key tasks in constructing environmental problems discussed in Chapter 5, trend spotters and theory testers can be said to characterise the 'assembling' process, communicators in 'presenting' an issue and applied policy analysts in 'contesting' an environmental claim.

Regulatory science and the environment

One important arena in which environmental science interacts with politics is in the regulatory process. The 'regulatory science' that is found here differs from conventional research science in a number of ways (Jasanoff 1990). First, it is done at the margins of existing knowledge where fixed guidelines for evaluation may often be unavailable. Second, it usually involves a greater degree of 'knowledge synthesis' than does research science, which puts a greater emphasis on the originality of findings. Third, science-based regulation requires a hefty dose of 'prediction' especially with regard to risk creation.

Jasanoff (1990: 230) argues that a negotiated and constructed model of scientific knowledge 'closely captures the realities of regulatory science'. Rather than encouraging an adversarial process, regulatory agencies seek scientific input into their decisions as a means of legitimation. This often takes the form of an ongoing scientific advisory committee. Jasanoff reviews a number of cases in which such advisory boards played a key role in decision-making at the Environmental Protection Agency (EPA) in the United States. In the case of air pollution, the relationship between the EPA and the Clean Air Scientific Advisory Committee (CASAC) was initially rocky but after extensive negotiation was transformed into a fundamentally sympathetic orientation. Similarly, despite problems during the Reagan era, the EPS's agency-wide Scientific Advisory Board (SAB) was able to maintain a respected and autonomous position, in large part because it focused on issues pertaining to scientific assessment while leaving rule-making activities to the agency proper.

In this negotiated model of science, Jasanoff contends, there can be no 'perfect, objectively verifiable truth', only a 'serviceable truth' which balances scientific acceptability with the public interest. In this context, scientific reality is clearly socially constructed so as to conform to a societal mean. However, in circumstances where sharply conflicting constructions of science land at the feet of a scientific advisory committee, reconciliation can often be most difficult. This is what has occurred in various regulatory controversies involving agricultural pesticides where scientific evidence has been especially difficult to establish while public concern has been high. In these situations, the debate over the 'precautionary principle' which we surveyed earlier in this chapter rears its head, with scientific advisers opting for the traditional reductionist position while agency staff are more sensitive to the public pressure to act sooner rather than later. Where this occurs, the risk debate can easily shift to the arenas of the media and politics where it will continue under a different set of ground rules from those encountered in the regulatory setting (Jasanoff 1990: 151).

8 Risk

To an ageing population concerned about preventing heart disease, salmon has proven to be a tasty remedy, especially in summer when it can be grilled to perfection on the barbecue. Thanks to the worldwide growth of aquaculture, consumers can obtain farmed salmon year-round at relatively low prices. Eating 'oily' fish such as salmon twice a week, the American Heart Association tells us, confers the health benefits of Omega 3 fatty acids that can reduce the risk of sudden cardiac death following a heart attack.

Suddenly, in January 2004, all bets seemed to be off when the respected journal *Science* published an article warning that farmed salmon contains alarmingly high levels of cancer-causing toxins, ten times more than in wild salmon. Risk analysis indicates, the authors warned, that 'consumption of farmed Atlantic salmon may pose health risks that detract from the beneficial effects of fish consumption' (Hites *et al.* 2004: 226).

As it happens, the *Science* piece was not the first research to come up with results of this type. Three years before, BBC News broadcast a programme, *Warnings from the Wild, The Price of Salmon*, that reported the results of a pilot project conducted under the auspices of the David Suzuki Foundation, that found farmed salmon had a higher level of PCBs and two others toxins than did wild salmon. Then, in July 2003, the Environmental Working Group (EWG) in Washington released a report entitled 'PCBs in Farmed Salmon: Factory Methods, Unnatural Results'. The EWG bought salmon from local grocery stores in the US. When the samples were analysed in the lab, it was found that seven out of ten fish were seriously contaminated with PCBs, raising concerns about cancer and foetal brain development. Based on their data analysis, the EWG concluded that 800,000 American adults ingest enough PCBs from farmed salmon to exceed the allowable lifetime cancer risk 100 times over.

Other health agencies and researchers hopped on the defensive. The US Food and Drug Adminstration (FDA) advised that the levels of pollutants found in salmon are too low for serious concern and urged Americans not to let the new research frighten them into a diet change. Eric Rimm, a specialist on nutrition and chronic disease at the Harvard School of Public Health, told the Associated Press that the *Science* article 'will likely over-alarm people in this country' (CNN January 9, 2004, http://edition.cnn.com/2004/HEALTH/01/08/salmon.pollution.) It was pointed out that the study tested salmon raw with the skin on – removing it and grilling the fish removed a significant amount of PCBs, dioxins and other pollutants. One university toxicologist, an industry consultant, went so far as to venture that 'in my view, the study says we should be eating more farmed salmon' (Stokstad 2004: 154).

Despite this counter-offensive, salmon as a healthy meal choice had lost its lustre. Some consumers began to avoid salmon altogether. Others insisted on wild salmon, a questionable strategy since many stores and restaurants routinely sell farm-raised salmon as 'wild' (Burros 2005). At a restaurant dinner with sociology colleagues shortly after publication of the *Science* article, none of those present would even consider ordering the salmon, citing recent research that stated that this was 'risky'.

To a large extent, this episode is characteristic of how individuals in contemporary society engage in the processes of risk perception and assessment. Typically, we hear a brief item on the radio or see it in a newspaper or on the Internet, it comes from a seemingly reputable scientific source and it taps into an existing well of concern about our health or the safety of our family. This is true not only for food and lifestyle choices but also for risks related to technology and the natural environment.

Until recently, the published literature on risk almost uniformly reflected the belief that risks should be 'objectively' determined, that this determination was exclusively the province of engineers, scientists and other experts and that any failure on the part of ordinary citizens fully to accept this was considered irrational. Risk assessment was thus conceived of as a technical activity where results were to be formulated in terms of 'probabilities'. There was even an emerging category of specialists – what Dietz and Rycroft (1987) have termed the 'risk professionals' – who make it their business to work out new methods of risk analysis.

Risk and culture

The first notable challenge to this position came from a British social anthropologist, Mary Douglas, and an American political scientist, Aaron Wildavsky, who published a provocative book in 1982 entitled *Risk and Culture: An Essay on the Selection of Technological and Environmental Dangers*.

Risk and Culture asks two simple but fundamental questions. Why do people emphasise certain risks while ignoring others? And, more specifically, why have so many people in our society singled out pollution as a source of concern? The answers, Douglas and Wildavsky insist, are embedded in culture.

In their view, social relations are organised into three major patterns: the individualist, the hierarchical and the egalitarian. Individualist arrangements are based on the laws of the marketplace while hierarchical relations are epitomised by government bureaucracies. Egalitarian groups are aligned in a 'border zone' on the margins of power at the political economic centre of society where the other two modes of social organisation are normally located.

Egalitarian groups have a cosmology or world-view that is more or less equivalent to the 'New Ecological Paradigm' discussed by Catton and Dunlap. Unbridled economic growth is frowned upon, the authority of science is questioned and our boundless faith in technology is declared unwise.

Douglas and Wildavsky's central thesis is that the perception of risk varies considerably across these three forms of social organisation. Market individualists are primarily concerned with the upswing/downturn of the stock market, hierarchists with threats to domestic law and order or the international balance of power and egalitarians with the state of the environment. This leads them to conclude that the selection of risks for public attention is based less on the depth of scientific evidence or on the likelihood of danger but rather according to whose voice predominates in the evaluation and processing of information about hazardous issues.

In this view, the public perception of risk and its acceptable levels are 'collective constructs' (Douglas and Wildavsky 1982: 186). No one definition of risk is inherently correct; all are biased since competing claims, each arising from different cultures, 'confer different meanings on situations, events, objects, and especially relationships' (Dake 1992: 27). Here, they are making the important point that competing definitions of what is risky are ultimately moral judgments about the proper way to organise society (Kroll-Smith *et al.* 1997: 8).

Unfortunately, at this point, Douglas and Wildavsky's cultural theory of risk slips off the rails on to spongier ground. Environmental egalitarians, they suggest, are the secular equivalents of religious sects such as the Anabaptists, the Hutterites and the Amish. Obsessed with doctrinal purity and the need for unquestioned internal loyalty, sectarians are seen as having to create an image of threatening evil on a cosmic scale. It is therefore necessary and 'functional' for environmental sectarians such as those found in Friends of the Earth constantly to identify new risks ranging from nuclear winter to global warming. Each new crisis is chosen, they maintain, 'out of the necessity of maintaining cohesion by validating both the sect's distrust of the center and its apocalyptic expectations' (Rubin 1994: 236). This explains why they turn their back on local causes in favour of global issues so vast in scale as to warrant a sense of general doom. Pollution and other risks are wielded by these sectarian challengers as a way of holding their membership together and for attacking the establishment groups of the centre, which they oppose (Covello and Johnson 1987: x).

Risk and Culture has provoked much interest and a torrent of criticism. Much of the latter focused on the authors' claim that environmentalists mobilise for solidary rather than for purposive reasons. That is, rather than view environmentalism as part of a moral response to a very real societal crisis, they have chosen to treat risks as merely bogeymen which serve the same purpose as certain food prohibitions among tribal peoples. Environmentalists, therefore, are not regarded as rational actors but rather as 'true believers' open to manipulation by ecological prophets such as David Brower and Edward Abbey.

Karl Dake, a member of the Douglas-Wildavsky research circle, has insisted that this criticism is overstated and that the cultural school of risk never meant to imply that perceived dangers are simply manufactured:

> People do die; plant and animal species are lost forever. Rather, the point is that world views provide powerful cultural lenses, magnifying one danger, obscuring another threat, selecting others for minimal attention or even disregard.
>
> (Dake 1992: 33)

Douglas and Wildavsky are less accommodating, however, insisting that knowledge about risk and the environment is 'not so much like a building eventually to be finished but more like an airport always under construction' (1982: 192). It is fruitless, they claim, for the social analyst to try to assess whether the risk under discussion is real or not; what matters is that the debate keeps going 'with new definitions and solutions'. Rubin (1994:

238–9) totally rejects this relativism, arguing that public policy considerations require that we know *definitively* whether risks such as those arising from global warming or ozone depletion are merely foils for the apocalyptic needs of sectarian organisations or genuine threats which must be dealt with. While Rubin's point is well taken, the ambiguity of many contemporary risks makes it difficult to achieve the certainty that he would like to see. Even if we reject Douglas and Wildavsky's absolute relativism, nevertheless, the by now widely accepted argument that they make about the subjective and imprecise nature of scientific findings militates against the infallibility of expert opinion. As a society, we still have to make social judgements about the magnitude of risk, although scientific evidence can be one helpful source of information in making these decisions.

Wilkinson (2001) has highlighted the similarities and differences between Mary Douglas and Ulrich Beck, whose 'risk society' thesis we examined in Chapter 2. Between them, he observes, 'they have provided the most detailed theoretical explanations for the social development of a new culture and politics of risk' (p. 1). Both theorists have chosen to address risk on a societal scale. Both point to the cultural relativity of risk perception and use the arguments of social constructionism. Neither is tempted to investigate empirically the prevalence of risk or the nature of risk perception. However, they differ as to the 'reality' of the risks we face. As we have seen, Beck embraces an apocalyptic vision of the future that is assured unless we engage in a new process of collaboration and social learning. By contrast, Douglas 'would cast doubt on the credibility of such an alarmist scenario and prefers to entrust herself to the professional opinion of government experts' (ibid).

Sociological perspectives on risk

Sociologists of risk generally adopt a more moderate position than that of Douglas and Wildavsky, insisting that while risk is certainly a sociocultural construct, it cannot be confined to perceptions and social constructions alone. Rather, technical risk analyses are an integral part of the social processing of risk (Renn 1992).

Dietz *et al.* (2002) observed, in preparatory work, that the main currents in the sociology of risk have followed three separate but complementary directions which are bound together by an underlying emphasis on the social context in which individual and institutional decisions about risks are made.

First, sociologists have been concerned with the question of how perceptions of risk differ across populations facing different life chances and whether the framing of choices stems primarily from power differences among social actors. Thus, Heimer (1988) points out that the residents of Love Canal saw the risks from chemical dumps differently from executives of the Hooker Chemical Company and from bureaucrats in the state government and various state agencies which deal with public health and the environment. Similarly, workers and bosses see environmental health risks in the workplace in a different light. To a certain extent, this issue overlaps the social distribution of risk, although the emphasis here is on how social location affects the perception of risk rather than on how it alters the likelihood of being exposed to hazardous conditions.

Second, sociologists of risk have proposed a model that reconceptualises the problem of risk perception by taking into account the social context in which human perceptions are formed. That is, individual perception is powerfully affected by a panoply of primary influences (friends, family, co-workers) and secondary influences (public figures, mass media) which function as filters in the diffusion of information in the community. This is captured in the concept of 'personal influence' that was central in the mass communication research of the 1950s and 1960s (see Katz and Lazarsfeld 1955).

Third, risks, especially those of technological origin, have been conceptualised as components of complex organisational systems. This is exemplified in Perrow's (1984) analysis of 'normal accidents' in which an estimated probability of failure is built right into the design of technologies with high catastrophic potential. Once implemented, however, such systems severely limit any further human ability to manipulate risks since the source of the risk is now located in the organisation itself (Clarke and Short 1993).

Renn (1992) has further classified the sociological approaches along two dimensions: (1) individualistic versus structural, and (2) objective versus constructionist. The first dimension asks whether the approach in question maintains that the risk can be explained by individual intentions or by organisational arrangements. Objectivist concepts imply that risks and their manifestations are real, observable events while constructionist concepts claim that they are *social artifacts* fabricated by social groups or institutions. According to this taxonomy, the first two currents of risk research identified by Dietz and his colleagues tend to be individualist/constructionist while the third is structural/objective. Notable by its absence is a 'social constructionist' perspective that Renn describes as an approach that 'treats risk as social constructs that are determined by structural forces on society' (1992: 71).

Social definition of risk

Hilgartner (1992) has argued that the constructionist perspective must begin by examining the conceptual structure of social definitions of risk. Such definitions, he maintains, include three major conceptual elements: an object deemed to pose the risk; a putative *harm*; and a *linkage* alleging some causal relationship between the object and the harm.

To assume that objects are simply waiting in the world to be perceived or defined as risky is 'fundamentally unsociological' (Hilgartner 1992: 41). Rather, an initial phase of risk construction consists of isolating and targeting the object(s) that constitute(s) the primary source of a risk.

In the late 1980s, the lakeside Toronto neighbourhood in which my family and I resided was designated by the municipal public works department to receive a pair of 'sewage detention tanks', one to be installed in Kew Gardens, a multi-use community park, the other on the beach adjacent to the boardwalk. The problem, we were told, was effluent from the City's storm sewer system that flowed into Lake Ontario and made it too polluted with faecal coliform bacteria to allow swimming. According to studies conducted by an engineering firm engaged by the City, there were two primary sources from which the faecal coliform pollution originated: human faeces contained in combined sewer overflow[1] and animal excrement which had been swept along with rainwater into the storm sewers.

Our residents' association, which first learned of the project when one member came across the publication of a statutory notice buried in the pages of a local daily newspaper, at first expressed concern on the grounds of the disruption which construction would bring to the park and the beach, both of which are heavily used. However, in the course of researching the proposal and meeting with other residents, we began to realise that, in fact, the source of the risk probably did not reside primarily in the storm-water but in effluent which was being dumped into the lake from the main sewage treatment plant located just to the west of our neighbourhood. We learned that, due to insufficient capacity, operators at this plant routinely opened the sea-wall gates just before it began to rain and released untreated or partially treated sewage into the lake at levels 10,000 times that at which the beaches were declared unsafe for swimming and closed. On one day out of three the lake currents reversed direction, sending this effluent towards our beaches. Immediately after a public meeting one night, a retired operator at the drinking-water filtration plant located at the eastern fringe of the neighbourhood told me that he used routinely to receive a telephone call from his equivalent at the sewage treatment plant advising that in advance of rain they were opening the gates and that he should raise the chlorine levels – a tip-off that the coliform pollution was migrating along the near shore area in a kind of bathtub ring pattern. We did not know it at the time but a somewhat similar situation occurs regularly in Sydney, Australia, where the ageing sewage system which pumps sewage out to sea is designed to overflow into storm sewers during periods of heavy rainfall so as not to clog up already overloaded treatment tanks (Perry 1994: WS–4).

What happened here was that residents opposed to the sewage detention tanks developed an alternative definition of the 'risk object'. At public meetings, at City Hall and at a special hearing before an Environmental Assessment Advisory Committee appointed by the Provincial Minister of the Environment to consider whether to grant our request for a 'bump up' (i.e. from a routine class environmental assessment to a more formal and rigorous individual environmental assessment), we actively contested the official designation of the object deemed to be risky and presented our claim (unsuccessfully) that the main sewage treatment plant was the villain instead.

The second element in the social definition of risk involves the process of defining harm. Once again, this is not as obvious as it may seem. For example, forest fires are commonly thought to wreak a path of destruction but ecologists contend that in nature they serve a useful function in woodland renewal. Offshore oil-drilling platforms are assumed to pollute the waters surrounding them but marine biologists have found that they also spawn a whole new micro-ecology at their base. Some environmentalists in the United States have campaigned to reduce allowable levels of the trace mineral selenium which can be added to animal rations on the basis that it leaves toxic residues, but representatives of the feed industry maintain that selenium additives are a boon to the environment because they reduce the amount of feed consumed thus saving on energy.

In each of these cases, the very definition of what harm ensues from a particular object or action is contested, sparking a variety of claims and counter-claims, despite the fact that there is mutual agreement as to the risk object (forest fires, offshore oil drilling, selenium as a feed additive). Risk claims may frequently conflict on ideational grounds. Thus, a river diversion project which provides irrigation water for local farmers (a human benefit) may result in the destruction of a fragile ecosystem of fish, birds, insects, etc. (a biological harm). Similarly, road salt, deemed so vital in order to cope with the harsh winter in parts of Canada and the Northern United States, has been labelled by scientists as harmful to the ecology of the lakes, rivers and streams where it is eventually deposited. Conversely, initiatives that are declared to be of ecological benefit may result in problems for human constituencies. For example, the protection of wolves is advocated by wildlife preservationists but it is keenly opposed by ranchers who fear the loss of livestock crucial to their economic survival. With consensus impossible, the central basis of contestation becomes the presence or absence of harm generated by a risk object.

A third component of the social construction of risk consists of the linkages alleging some form of causation between the risk object and the potential harm. Hilgartner (1992: 42) observes that constructing these linkages is always problematic because a risk can be attributed to multiple objects. Indeed, the 'laws' of ecology encourage this since all things are regarded as being interdependent. This is further complicated by the fact that the full extent of the risk may not be known until many years later. For example, a report in the mid-1990s by a Minnesota radio station suggested that a 1953 US Army test, in which clouds of zinc cadmium sulphide, a suspected carcinogen, were sprayed aerially over Minneapolis dozens of times have caused an unusual number of stillbirths and miscarriages; these problems have shown up particularly often in former students of a public elementary school which was one of the spray sites forty years before (*New York Times* 1994). The effects may sometimes be more immediate but it takes years for claims-makers to assemble them into a publicly acknowledged form. This has been the case with a raft of health problems among military veterans of the Gulf War. Even though symptoms began soon after their return, it took some time for public reports of a 'Gulf War syndrome' to penetrate the mainstream media and to be framed in terms of toxic environmental agents in the war zone.

Much of the discourse over the construction of risk takes place on this terrain. The situation is further complicated by the existence of multiple conflicting proofs: legal, scientific, moral.

The burden of legal proof is most onerous, since it cannot leave any room for 'reasonable doubt'. The caveats that are standard in scientific studies (e.g. 'the data are suggestive but require further research') do not stand up in court. Nor usually does anecdotal or clinical evidence.[2] As environmentalists have discovered, judges are often loath to break any new ground by acting to prevent a problem before it happens. As Freudenburg (1997: 34–5) pointed out, the capability of the courts to deal with technological risks and disasters is especially limited by 'the need to establish clear and unambiguous liability, even in the presence of evidence that will remain at best probabilistic'.

Scientific proof is easier to come by, but nevertheless is a slave to statistical levels of significance. It is also notoriously fickle, its authority intact only until the next disconfirming study appears. The scientific layer of proof can be subdivided into two: a standard drawn from pure science in which action is not recommended until correlations weigh in at the 95 per cent confidence level, and a standard utilised by the medical disciplines in which action may be taken before significance is reached if the evidence points towards a serious health problem.

Collingridge and Reeve (1986) have demonstrated the clash between these two versions of scientific evidence in the debate over the health effects on children of lead from vehicle exhausts. In the United States, it haunted the conflict between the EPA, which supported the removal of lead gasoline (petrol) on basis of broad differences in blood lead levels among urban and suburban populations, and the Ethyl Corporation, a major manufacturer of lead additives, which argued that the link between blood and air levels remained statistically unproven. In the UK, difficulties arose in early 1980s between the government-sponsored

'Lawther Report' which rejected all laboratory animal and biochemical studies as irrelevant to understanding the medical effects of lead on humans and the report entitled *Lead or Health* by the environmental group, the Conservation Society, which argued the contrary: 'Moral proofs are most easily manufactured but are heavily dependent upon the mobilisation of public opinion in order to make an impact.'

The use of moral proofs allows the formation of attitudes or opinions about a risk issue even if the scientific or legal layers of proof indicate a degree of uncertainty or ambiguity. For example, animal rightists have never been able to prove conclusively that animals 'suffer' so they have adopted the alternative strategy of trying to demonstrate ethically that this is the case, drawing in particular on the work of the philosopher Peter Singer. Similarly, the scientific case against the biological engineering of plants and animals is still inconclusive (no genetically altered fruits have thus far performed like the protagonist in the Roald Dahl story, *James and the Giant Peach*) but the moral case against interfering with nature is more impressive. Such moralisation, however, tends to polarise positions on risk policies, making compromises more difficult (Renn 1992: 192).

Unlike the legal and the scientific, the most effective moral proofs are often those that follow a simple line of reasoning. Consider, for example, the nature of the argument presented by 'Kapox' – labelled by the South American press as the 'Tarzan of the Amazon'. Kapox, who engages in long-distance swims through the Amazon region to publicise the state of pollution of the river and the destruction of the surrounding rainforest, does not base his appeal on a sophisticated reasoning about the need to protect biodiversity. Rather, he preaches a simple, obvious, moral message: as the largest river in the world concentrating a fifth of the planet's fresh water, the Amazon deserves respect (Suzuki 1994a).

Arenas of risk construction

As powerful as Kapox's appeal may be, it is unlikely to influence collective risk decisions or policies directly. Instead, social definitions of environmental risk must be followed up by political actions designed to mitigate or control the risk that has been identified. Building on the work of Hilgartner and Bosk (1988), Renn (1992) argues that political debates about risk issues are invariably conducted within the framework of 'social arenas'.

The term *social arenas* is a metaphor to describe the political setting in which actors direct their claims to decision-makers in hopes of influencing the policy process. Renn conceives of several different (theatre) 'stages' sharing this arena: legislative, administrative, judicial, scientific and mass media. While both traditional and unorthodox action strategies are permitted, these arenas are nevertheless regulated by an established repertoire of norms. For example, illegal direct action such as that advocated by Earth First, the American renegade environmental group, violates this protocol. The code is, in fact, a combination of formal and informal rules usually monitored and coordinated by some type of enforcement or regulatory agency such as the EPA in the United States and the Department of the Environment (DoE) in Britain.

The concept of the social arena combines elements from the organisation–environment perspective in the field of complex organisations, Goffman's dramaturgical model of social relations and the symbolic models of politics as developed by Murray Edelman (1964; 1977) cemented together by a social constructionist compound. As formulated by Renn, it also stresses the mobilization of social resources as discussed by the McCarthy–Zald school within the resource mobilisation perspective on social movements. Renn seems unaware of the parallels but the social arena concepts that he uses also echo some basic research on international environmental diplomacy, notably Haas's (1990; 1992) seminal concept of 'epistemic communities'.

While some elements of risk construction may occur in the public domain beyond their parameters, the most important action takes place in arenas that are populated by communities of specialised professionals: scientists, engineers, lawyers, medical doctors, corporate managers, political operatives, etc. (Hilgartner 1992: 52). Such technical experts are the chief constructors of risk, setting an agenda that often includes direct public input only during the latter stages of consideration. Hilgartner and Bosk (1988) note that these 'communities of operatives' often function in a symbiotic fashion, the operatives in each arena feeding the activities of operatives in the others. Environmental operatives (environmental groups, industry lobbyists and public relations personnel, political champions, environmental lawyers, journalists and bureaucrats) are notable examples of this; by virtue of their activities they both generate work for one another and raise the prominence of the environment as a source of social problems.

Within the social arena of risk, the process of defining what is acceptable and what is not is often rooted in negotiations among several or multiple organisations seeking to structure relations among themselves. Clarke (1988) illustrates this in his analysis of an office building fire in Binghampton, New York, which left a legacy of toxic chemical contamination. In this case, three governmental agencies – the state health department, the county health department and the state maintenance organisation – collectively vied for suzerainty in determining how risky the situation was thought to be. In such cases, Clarke argues, the institutional assessment of risk is a claims-making activity in which corporations and agencies both compete and negotiate to set a definition of acceptable risk.

From a theatrical vantage point, social arenas of risk are populated by sundry groups of actors. Palmlund (1992) proposes the existence of six 'generic roles' in the societal evaluation of risk, each of which carries its own dramatic label: risk bearers, risk bearers' advocates, risk generators, risk researchers, risk arbiters and risk informers.

Risk bearers are victims who bear the direct costs of living and working in hazardous settings. In the past, those who are impacted most have rarely asserted themselves and have therefore remained on the margins of risk arenas. More recently, however, as can be seen in the rise of the environmental justice movement, risk bearers have become empowered and must increasingly be regarded as notable players. *Risk bearers' advocates* ascend the public stage to fight for the rights of victims. Examples include consumer organisations such as those headed by Ralph Nader and Jeremy Rifkin, health organisations, labour unions and congressional/parliamentary champions. They are depicted as protagonists or heroes. *Risk generators* – utilities, forestry companies, multinational chemical and pharmaceutical companies, etc. – are labelled as antagonists or villains since they are said by advocates to be the primary source of the risk. *Risk researchers*, notably scientists in universities, government laboratories and publicly funded agencies are portrayed as 'helpers' attempting to gather evidence on why, how and under what circumstances an object or activity is risk-laden, who is exposed to the risk and when the risk may be regarded as 'acceptable'. On occasion, however, risk researchers have become identified with risk generators, particularly if their findings support the latter's position. *Risk arbiters* (mediators, the courts, Congress/Parliament, regulatory agencies) ideally stand off-stage seeking to determine in a neutral fashion the extent to which risk should be accepted or how it should be limited or prevented and what compensation should be given to those who have suffered harm from a situation judged to be hazardous. In reality, risk arbiters are rarely as neutral as they should be; instead, they frequently they tend to side with risk generators. Finally, *risk informers*, primarily the mass media, take the role of a 'chorus' or messengers, placing issues on the public agenda and scrutinising the action.

Renn (1992) suggests a hybrid of several of these roles: the *issue amplifiers* who observe actions on stage, communicate with the principal actors, interpret their findings and report them to the audience. Environmental popularisers such as Paul Ehrlich, Barry Commoner, Jeremy Rifkin and Jonathan Porritt are prime examples of this.

Hilgartner and Bosk depict the interaction among different arenas of public discourse as characterised by several key features. First, these multiple arenas are connected by a complex set of linkages both social and organisational. As a result, activities in each arena thoroughly propagate throughout the others. Second, one finds a huge number of 'feedback loops' that either amplify or dampen the attention given to problems in public arenas. Consequently, you find a relatively small number of successful social problems that occupy much of the space in most of the arenas at the same time. This synergistic pattern is typical of policy-making on matters relating to risk and the environment.

In their study of 228 Washington-based 'risk professionals', Dietz and Rycroft (1987) found a policy community with a dense network of communication which stretched across environmental groups, think-tanks, universities, law and consulting firms, corporations and trade associations, the EPA and other executive agencies. Environmental organisations were especially active in outreach activities including contacts with corporations and trade associations with whom 85 per cent of respondents communicated in a typical month. Similarly, the personnel flows across organisations, another component of the exchange network, was substantive, although working for an environmental group led to a low probability of finding employment with one of the other groups.

Dietz and Rycroft depict the environmental risk policy system as a hybrid in the sense that it has a strong base in science but at the same time is driven by the ideological conflict between environmentalists and corporate and governmental participants. This creates a measure of volatility insomuch as science is the cornerstone of the system yet many key decisions are resolvable only in political terms. Nevertheless, the picture that emerges from this survey study is one of a policy community that is permeable but nevertheless closely linked and oriented towards a shared discourse on issues relating to environmental risk. Among other things, this means that any approach to risk that attempts to emphasise sociocultural facts over material ones will probably be considered off target and therefore inappropriate for inclusion on the shared agenda of risk professionals (Dietz and Rycroft 1987: 114).

Power and the social construction of environmental risk

Freudenburg and Pastor (1992) have observed that the social constructionist approach to risk is well positioned to discuss risk construction in the context of power. In a similar fashion, Clarke and Short (1993) note that constructionist arguments – in contrast to those anchored in psychology and economics – tend to focus on how power works in framing terms of debate about risk.

Both sets of authors share the belief that this relationship is especially important because official viewpoints, with their significantly greater access to the mass media, strongly suggest that public fears regarding technical risks are clearly irrational; that is, claims about public irrationality are in themselves ways of framing risk issues. By implication, policy formulations that originate with the community of risk professionals (see the previous section) are presented as rational, objective assessments of what is considered safe and what is not. If this view is accepted, then the central task is said to be educating the public to realise that they are over-reacting and that nuclear power, herbicides, bioengineered organisms, etc., are not really the hazards that they appear to be. In order to allay public fears, risk analysts develop quantitative measures through which to compare the risks inherent in different policy choices and their relative costs and benefits (Nelkin 1989: 99).

This is not to imply that the people are always right and the knowledge of the experts invariably 'brittle' (Wynne 1992). Rather, a social constructionist perspective would argue that each represents a competing frame but the dominant rationality that comes from the risk establishment is superimposed over the popular frame due to a power differential. Thus, Wynne (1992: 286) demonstrates in the case of a public controversy over the herbicide 2,4,5-T in the United Kingdom that the firsthand empirical knowledge of farm and forestry workers was directly relevant to an objective risk analysis. However, scientists flatly refused to consider this knowledge as legitimate, thereby denigrating and threatening the social identity of the local citizens.

Nowhere is this differential more evident than at public information meetings or hearings that are routinely stage-managed by risk generators and arbiters. At the public meetings concerning the building of the sewage detention tanks described earlier in this chapter, members of the public works department, local politicians (who strongly supported the project) and representatives of the private engineering firm who had recommended the building of the tanks all sat together on the elevated stage of the auditorium whose perimeters were adorned with charts, blown-up photographs and other 'props'. We citizens were restricted to a single question with no follow-up. Those who queried the suitability of the project were alternately bullied and patronised. On contentious issues the presenters did not hesitate to introduce a ream of previously unseen statistical evidence that we had no way of confirming or denying without days or weeks of further research.

Richardson *et al.* (1993) observed many of the same structural elements in the conduct of a set of environmental public hearings in 1984 on the proposed building of a bleached kraft pulp mill in Northern Alberta, Canada.[3] The members of the Alpac EIA Review Board who were conducting the hearing sat at a table facing the public on a stage. At one of several tables to the direct right of the Board were the representatives of Alberta–Pacific Forest Industries (Alpac), the company that sought to build the mill, their technical experts and their lawyer. Numerous Alpac consultants were scattered throughout the proceedings. Presenters were required to speak into microphones through which their words were recorded.

Kaminstein (1988) argues that embodied in the public presentation of scientific information at meetings concerning the health and safety aspects of toxic waste dumps is a *rhetoric of containment* which restricts discussion, avoids tough questions and pursues its own agenda. Drawing on three years of observation of meetings held to inform residents of Pitman, New Jersey, about the steps which were being taken to clean up the Lipari landfill, the site of one of the worst dumps in the United States, Kaminstein concludes that residents were not so much informed or persuaded as controlled and defeated. The primary tool that scientific experts associated with the EPA and the Centers for Disease Control used to stifle citizen initiatives was *toxic talk* – talk that stifles discussion and smothers public concern. The rhetoric of containment has multiple elements.

First, as was the case with the detention tank meetings, residents were bombarded with technical information. At one meeting, EPA officials distributed documents totaling 44 pages. Those in attendance were expected to assimilate an array of data, charts, graphs, tables and a slide show in rapid succession. At the same time, the facts that residents wanted were never available, and no explanation or interpretation was given as to the information that the consultant scientists presented.

The physical setting of the meeting room was also very similar to that experienced by those attending the detention tank sessions. At the front of the room was a large dais raised about two feet, a long table and nine large, high-backed chairs on which the scientists sat, creating a physical and psychological distance from the audience. Various dramatic props, for example, an enlarged photograph of an air-monitoring vehicle that looked like a recreational camper, were employed as rhetorical devices to pacify the residents and enhance the power of those in charge of the meeting.

The factual presentation style used by EPA officials and scientists was abstract, impersonal and technical, thus creating an impression of professional neutrality. It was the activist residents who became angry and confrontational, allowing officials to dismiss them as 'emotional'. Questions that dealt with the geology and hydrology of the area, future tests and plans for the clean-up were addressed but those which dealt with health risks were avoided or deflected. Officials and scientists used language in their presentations that was technical, ambiguous and intellectual, making it impossible for any meaningful dialogue to develop between experts and residents over the nature and magnitude of the risks faced by the community of Pitman.

Toxic talk techniques such as this are strategically successful if ethically reprehensible. It allows scientific experts and government officials to direct the discussion, set the risk agenda and discourage future citizen participation. Popular concerns and risk frames are subordinated to those that are preferred by the powerful in society. As Kaminstein (1988: 10) notes, these kinds of exclusionary devices permit agencies such as the EPA legally to fulfil their mandate to hold public meetings while at the same time leaving residents feeling that they are fighting a losing battle just to be heard.

That is not to say that members of the public never attempt to assert themselves in settings such as these. For example, in the Alberta case, some participants fought to wrest control from regulators over the scope of the review, the venues and over definitions of legitimacy, as well as attempting to subvert the dominant discourse that was imposed by the pro-development forces (Richardson *et al.* 1993: 47). However, the constraints of the hearing process normally make effective citizen participation difficult, especially since the situation is structured so as to prevent public argument and reinforce the power of institutions.

Institutional risk analysts and regulators also exercise power on a broader plane. Structurally, they control the official risk agenda, acting as gatekeepers who are well placed to determine which issues are included or excluded from public discourse. For example, in the 1980s, imbued with the deregulatory climate within the Reagan administration (supported by senior EPA managers), Congress fatally slashed the budget of the Office of Noise Abatement and Control (ONAC) thereby also dooming most state and local noise abatement programmes (Shapiro 1993). Despite the continued risk posed by noise pollution to human health and environmental aesthetics, the issue stalled for lack of government action. In such circumstances, the risk itself does not diminish (in the case of noise pollution it in fact increased) but the risk establishment is able to manipulate its progress on the action agenda.

Freudenburg and Pastor (1992: 403) note that the social constructionist approach to technological risks would do well to look at other variables that sociologists have previously found to be associated with power. Thus gender may be significant here, insomuch as the scientific experts and bureaucratic officials who practise the rhetoric of containment are usually men while local citizen groups are disproportionately composed of women, many of whom lack power and authority in public life. Similarly, members of racial and ethnic minorities are routinely dismissed and discredited by the risk establishment, an experience that has led to the blossoming of the environmental justice movement. The relationship between power, inequality and the social construction of risk is equally evident in communities that have been marginalised by positions of economic, geographic or social isolation (Blowers *et al.* 1991).

Risk construction in cross-national perspective

Finally, risk construction varies cross-nationally according to a number of different factors: the organisation of political and administrative structures, historical traditions and cultural beliefs. Within the field of risk analysis, a classic comparative study is Sheila Jasanoff's (1986) report entitled *Risk Management and Political Culture*. Drawing on case studies of national programmes for controlling carcinogens in several European countries, Canada and the United States, she concludes that cultural factors strongly influence goals and priorities in risk management. In Germany, the favoured approach has been to delegate resolution of all risk-related issues to technical experts. Jasanoff does not discuss it but even where a risk subject is strongly contested, technical rationality is applied in the form of a 'technology assessment' that includes representatives from government, industry and social movements (see Bora and Dobert 1992). In Britain and Canada, risks are examined through a mixed scientific and administrative process but scientific uncertainties are not always publicly broadcast. By contrast, in the United States risk determination has a much more public face surfacing in a wide variety of administrative and scientific fora. While this can produce greater analytical rigour and more democratic and informed public participation, it can also lead to more polarisation and conflict and thus to political stalemate.

Using the comparative method suggested by Jasanoff, Harrison and Hoberg (1994) compared government regulation in Canada and the United States of seven controversial substances suspected of causing cancer in humans: the pesticides alar and alachlor, urea-formaldehyde foam insulation, radon gas, dioxin, saccharin and asbestos. Each country's approach was weighed according to five criteria for effectiveness: stringency; and timeliness of the regulatory decision; balancing of risks and benefits by decision-makers; opportunities for public participation; and the interpretation of science in regulatory decision-making.

As had Jasanoff, the researchers found that there were two contrasting regulatory styles. In each case:

> there was more open conflict over risks in the United States than Canada, with interest groups, the media, legislators and the courts playing a much more important role south of the border. The regulatory process in Canada tended to be closed, informal, and consensual, in comparison with the open, legalistic, and adversarial style of the US.
>
> (Harrison and Hoberg 1994: 168)

Both styles are said to have risks and benefits. The Canadian system is more conducive to scientific caution and formal democratic control but it lacks accountability, making it easier for political decisions to be cloaked in scientific arguments. The American system is more open but also more conflictual and vulnerable to interest group pressures and, as a result, less dependent upon scientific expertise.

This comparative research provides further evidence that risk determination and assessment are socially constructed. National political structures and styles can be seen to have as much to do with deciding which environmental conditions will be judged to be risky and actionable as the nature of the scientific claim itself. Consequently, fundamentally sound environmental claims may be deflected or stalled, either due to collusion between regulators and scientists or because of political pressure from interest groups, either within or opposed to the environmentalist perspective.

9 Biodiversity loss

The successful 'career' of a global environmental problem

Along with global warming, the conservation of biodiversity was one of the two major issues at the June 1992 United Nations Conference on Environment and Development (UNCED) in Rio de Janeiro. It was called the 'hottest' environmental topic of 1993 (Mannion 1993) with a burgeoning academic and popular literature devoted to exploring its parameters. Valiverronen (1999: 404) characterises it as 'the latest "big" environmental issue, comparable to acid rain, ozone depletion and climate change'. Yet twenty years before, the term biodiversity was unknown and it was not to be found in any compendium of threats to the environment. The skyrocketing career of biological diversity loss is a good illustration of how a 'transnational epistemic community' (see Chapter 7) can assemble, present and successfully contest a global environmental problem.

As a concept, biodiversity is multi-layered with various levels of meaning (Udall 1991: 82). Officially, it has been defined as 'the variability among living organisms from all sources, including *inter alia*, terrestrial, marine and other aquatic ecosystems and the ecological complexes of which they are part' (Tolba and El-Kholy 1992). More simply, it is an umbrella term for nature's variety – ecosystems, species and genes (Environmental Conservation 1993: 277).

Biodiversity is generally acknowledged to exist at three distinct levels: ecosystem diversity; species diversity; and genetic diversity.

Ecosystem diversity refers to the variety of habitats that host living organisms in a particular geographic region. This variety is said to be shrinking in the face of accelerating economic development. Udall (1991: 83) uses the metaphor of a ripe pumpkin that has been hollowed out to describe the damage to our ecosystems which has been inflicted by trapping, ploughing, logging, damming, poisoning and other forms of human intrusion. With the rapid pace of development, land ecosystems are described as increasingly taking the form of 'habitat islands'; for example, a patch of tropical forest surrounded by croplands (Franck and Brownstone 1992: 37).

Species diversity refers to the variety of species that are found in an ecosystem. While there have been notable episodes of species extinction in the past, the scale of loss today is judged to be unprecedented in the history of humankind (Lovejoy 1986: 16). Much of this is attributable to loss of ecosystem diversity; as a broad general rule, reducing the size of a habitat by 90 per cent will reduce the number of species that can be supported in the long run by 50 per cent (Tolba and El-Kholy 1992; 186).

Genetic diversity refers to the range of genetic information coded in the DNA of a single population species. Biologists value genetic diversity because it is seen as the basis for permitting organisms to adapt to environmental change. For example, in agriculture, wild strains of plants are valued because they often contain genes that are vital in fighting off pests or disease, unlike domesticated 'monocultures' which are much more vulnerable. In the animal world, inbreeding among a population stranded by habitat loss or commercial exploitation leads to an inability to survive in the long term; for example, this is the situation of the grizzly bears in Yellowstone Park in the American West (Udall 1991: 82).

When all three levels are viewed together, biodiversity loss appears to be a newly minted environmental problem. However, as Barton (1992: 773) has observed, there have long been a variety of treaties governing individual elements such as the international trade in endangered species, regional conservation and the conservation of particular species. For example, the Migratory Birds Convention, signed in 1917 by the United States and Canada, was a key piece of legislation in the campaign during the first part of this century to save North American birds. And in 1911, six years earlier, a major international agreement, the Convention for the Protection and Preservation of Fur Seals, had been signed.

Contextual factors

There are three major developments that set the stage for the rise of biodiversity loss as a major environmental problem in the 1980s and 1990s.

First, the growing economic importance of biotechnology meant that a greater financial value was increasingly being placed on genetic resources, a value that was recognised through intellectual property rights (Barton 1992: 773). Of special importance here was a landmark decision by the US Supreme Court (*Diamond v. Chakrabarty*) that allowed for the first time the patenting of a genetically engineered microbe, in this case an oil-eating bacterium developed by a General Electric research scientist named Ananda Chakrabarty. Also of significance was the passage a decade earlier of the US Plant Variety Protection Act (PVPA) that set up a patent-like system to govern the seed industry under the auspices of the US Department of Agriculture rather than under the more rigorous requirements of the US Patent Office. These events were significant for two interrelated reasons.

By raising the monetary stakes involved in the development of genetic resources, a conflict was fanned between the developed nations who wished to ensure open access to plant and animal genes and the less developed nations in which the bulk of these genetic materials were actually to be found. The latter began to see the genetic prospecting of the multinational pharmaceutical and chemical companies headquartered in Northern nations as a form of 'plundering' for which compensation should be paid.

At the same time, genetic diversity also became an international development issue due to the entry of several well-known rural activists (Cary Fowler, Pat Roy Mooney) to the debate over plant patenting. Fowler, a farmer from North Carolina, had worked with food activists Frances Moore Lappé and Joe Collins on the national bestselling book, *Food First*, an indictment of the world food system. Fowler became a one-person lobby opposing changes to the seed patent laws. In the 1979 debate over a proposal to amend the PVPA so as to add six 'soup vegetables' theretofore excluded from the act, Fowler:

> turned his mailing list loose on Congress, went to the Press, wrote articles about the issue, and travelled around the country alerting other groups to the 'seed patenting' issue. Fowler

> rallied scientists and church interests and wrote to the Secretary of Agriculture, Bob Bergland, urging him to consider the impact of rising seed costs on small farmers.
>
> (Doyle 1985: 67–8)

Mooney, a Canadian from the province of Manitoba, helped to internationalise the seed issue both by his participation in a network of activist scholars working on Third World issues and also through his widely circulated paperback book, *Seeds of the Earth*, published in 1979 for the Canadian Council for International Cooperation and the International Coalition for Development Action.

Second, the emergence of conservation biology in the late 1970s as an academic speciality provided a nesting spot for research on biodiversity. Conservation biology is an applied science that studies biodiversity and the dynamics of extinction. It differs from other natural resource fields such as wildlife management, fisheries and forestry by accenting ecology over economics (Grumbine 1992: 29). The role of the conservation biologist is to provide 'the intellectual and technological tools that will anticipate, prevent, minimize and/or repair ecological damage' (Soulé and Kohm 1989: 1). Conservation biology is thus a 'crisis discipline'[1] that draws its content and method from a broad range of fields within and outside of the biological sciences.

Conservation biology was formally recognised as a discipline in 1985 with the creation of the Society for Conservation Biology (SCB). Within three years, the membership of the Society had swollen to nearly 2,000 members (Tangley 1988: 444). SCB is significant because it has provided a central forum for the communication of knowledge about conservation and biological diversity, especially through its journal, *Conservation Biology*.

Another critical node in the development of the discipline was the establishment of the Center for Conservation Biology (CCB) in the Department of Biological Sciences at Stanford University in California. The Center's main activities are basic and applied research, education and the application of conservation biology principles to genetic resources, species, populations, habitats and ecosystems. CCB consults not only within the United States but also internationally, especially in Latin America (Franck and Brownstone 1992: 66) providing yet another link between research on biological diversity and the international development scene.

By the late 1980s, conservation biology had begun to develop rapidly at institutions of higher learning. A pioneering textbook, *Conservation Biology: Science of Scarcity and Diversity*, had been adopted by classes at 37 US colleges and universities as well as overseas (Tangley 1988: 444). In 1985, the first conservation biology course was taught at the University of California at Berkeley with an emphasis on the biological foundations for conservation (Millar and Ford 1988: 456). While research funding was still modest, partly because of a perception that conservation biology was a 'soft' discipline, advocates of biological diversity as an environmental problem nevertheless had an increasingly powerful academic medium for spreading their message and for building a constituency.[2]

Third, a legal and organisational infrastructure was being assembled in the 1970s within the United Nations and other NGOs[3] dealing with various elements of the biodiversity problem.

In 1971, the *Convention on Wetlands of International Importance Especially as Waterfowl Habitat* was agreed upon with the dual purpose of designating environmentally sensitive areas for migratory waterfowl and facilitating trans-border cooperation among countries situated along their travel routes. This agreement was staffed by a secretariat provided by the International Union for Conservation of Nature (IUCN).

The Convention Concerning the Protection of the World Cultural and Natural Heritage (held in Paris in 1972), prepared under UNESCO (United Nations Economic, Social and Cultural Organization) supervision, established exceptional World Cultural Sites such as Serengeti National Park in Tanzania, the Queensland Rainforests in Australia and Great Smokies National Park in the United States, some of which rated quite highly in biological diversity. The agreement established a world heritage fund to assist nations that may have difficulty in paying for the protection of these unique sites. It was signed by 150 countries. However, this treaty is extremely limited in scope and has had minimal success both in slowing the rate of species loss on a global scale and in assuring the protection of designated sites (Spray and McGlothlin 2003b: 154).

In 1973, the *Convention on International Trade in Endangered Species of Wild Fauna and Flora* (CITES) was proclaimed in Washington with a secretariat staffed by the United Nations Environment Programme (UNEP) located in Lausanne, Switzerland. This convention established lists of endangered species for which international trade is to be controlled via permit systems. CITES was limited, however, insofar as it was directed at individual species rather than the habitats in which they resided. Furthermore, it is primarily a trade agreement that does not guarantee protective status or conservation programmes within the states in which vulnerable species reside (Spray and McGlothlin 2003b: 155). Finally, the monitoring and enforcement of CITES has been marred by a series of 'exceptions', for example the 'tourist souvenir exception' (allows rare specimens to be imported as personal or household effects) and the 'trans-shipment exception' (permits specimens passing through a third country to avoid regulations of the convention). These exceptions have been used to smuggle protected species under the pretence that the exceptions apply (Louka 2002 :116–17).

Finally, the *Convention on Conservation of Migratory Species of Wild Animals* (CMS), also known as the *Bonn Convention* provided a framework for international cooperation among states that hosted animals whose travels regularly take them across national boundaries. A central aim of this convention was to coordinate research, management and conservation resources such as habitat protection and hunting regulation affecting migratory species.

These international legal agreements were supplemented by a number of regional measures, for example, conventions on the conservation of nature, natural resources, wildlife and natural habitats pertaining to the South Pacific (1976), Africa (1968) and Europe (1976), and by the designation of Biosphere Reserves under a UNESCO programme. Taken as a whole, such measures were not only useful in their own right as a means of fostering, if not enforcing, useful cooperation among nations in conserving biological diversity, but they also put into place a global system upon which more far-reaching and stringent international legislation to conserve biological diversity could be modelled. Furthermore, they established epistemic networks of research, communication and coordination that were vital in moving biodiversity along to its status today as a major environmental problem.

Assembling the claim

In contrast to those global environmental problems that involve damage by pollutants to the atmosphere (or stratosphere) – global warming, ozone depletion, acid rain – the threatened loss of biological diversity has been less dependent on the discovery of an alteration in nature; for example, the ozone 'hole' over the Antarctic or 'forest die-back' in the Black Forest. Rather, it has developed in the context of a steady outpouring of studies that have cumulatively rung the alarm bells.

Taken as a whole, these studies have often lacked precision, with the result that the projected number of extinctions that might be expected has varied not only widely, but also wildly (Brown 1986: 448). Estimates have frequently been made in terms of rates, a device that both implies a greater accuracy than is possible given current knowledge, leading to some questionable figures. Most notably, the 'one extinction per minute' rate used by some authors is equivalent to 525,600 extinctions per year, an unlikely or impossible total about ten times the number usually cited (Lovejoy 1986: 14). At the lower end, USAID (United States Agency for International Development) currently claims that 1,000 species per year are becoming extinct (www.usaid.gov/our_work/environment/biodiversity).

Furthermore, the enormity of the problem has meant that reliable information is difficult, if not impossible, to assemble. So little is actually known about how species interact in ecosystems, about how they depend upon each other and about how they recover from episodes of disturbance that 'actions required now to avoid future disasters must be undertaken without sufficient knowledge to make considered choices' (Norton 1986: 11).

Most current methodologies for the assessment of biodiversity use either of two methods: the measurement of species and the identification of genetic diversity. The former is inadequate insofar as it is not always the appropriate unit of measurement (use of phyla and families may be more accurate); it is not necessarily the best way of locating diverse ecosystems; and it does not provide for changes in species and habitats over time. Identification of genetic diversity is even more difficult, insofar as it is expensive, requires trained personnel capable of using sophisticated laboratory techniques, and produces difficult-to-interpret results (Louka 2002: 124).

Finally, some scientists have questioned whether existing efforts to quantify biodiversity loss rates are flawed because they incorrectly assume that extinctions are 'random'. Thus, Raffaelli (2004) has argued that in the real world most extinction events are non-random, that is, some species are more likely to go extinct than others. Such non-random extinctions may have greater consequences for species loss than those predicted on the basis of studies in which extinctions are assumed to occur randomly (p. 1142).

In the face of this scientific uncertainty, those who have promoted biodiversity as an environmental problem have fallen back on the 'precautionary principle' (see Chapter 7) suggesting that the wisest course is simply to avoid actions that needlessly reduce biological diversity (Tolba and El-Kholy 1992: 197).

How, then, were conservation biologists and other claims-makers able to elevate biodiversity loss to the status of a notable environmental problem, given a relative lack of authoritative research data on the subject?

Wilson (1986: v) believes that the rising interest during the 1980s among scientists and portions of the public in matters related to biodiversity and international conservation can be ascribed to two more or less independent developments.

The first was the convergence of data from three different areas of research – forestation, species extinction and tropical biology – such that global biodiversity problems were brought into sharper focus and thought to warrant broader public exposure. This critical mass of data was not sufficient to build an airtight case for immediate worldwide action but it did raise the profile of biodiversity to a level sufficient to provoke a stream of academic conferences, political hearings and public forums.

The largest of these was the National Forum on BioDiversity held in Washington, DC, on 21–24 September 1986 under the auspices of the National Academy of Sciences and the Smithsonian Institution. This forum featured sixty leading biologists, economists, agronomists, philosophers and international development experts. The lectures and panels were regularly attended by hundreds of people and the final evening's panel was teleconferenced to an estimated audience of 5,000 to 10,000 at over a hundred universities and colleges in the United States and Canada. It was at this conference, Wilson (1986: vi) recalls, that the term 'biodiversity' was first introduced by the organiser Dr Walter G. Rosen, a programme officer from the Commission on Life Sciences, National Research Council/National Academy of Sciences. It is also worth noting that in spite of Wilson's protests that the term biodiversity was too 'catchy' and 'lacks dignity', Rosen and the other Academy staff members persisted on the grounds that the term is simpler and more distinctive, and therefore the public would remember it more easily (Wilson 1994: 359).

The second development was the growing awareness of the close link between the conservation of biodiversity and economic development, especially in Third World nations. This elevated biodiversity loss from a scientific environmental problem to a wider status as a sociopolitical problem. In the industrial nations of the North, destruction of tropical rainforests and other Third World habitats was decried on the basis that it threatened a vast untapped reservoir of species that could potentially prove useful in providing new foods, medical treatments and other products. At the same time, in the countries of the South, biodiversity loss was feared for its impact on local farmers and others whose livelihoods depend on the maintenance of traditional ecosystems. In time, these two objectives were to come into direct conflict, but initially they acted in concert so as to reframe biodiversity loss as a 'development' problem of considerable importance. According to USAID, the net economic benefits of biodiversity in 2005 are estimated to total at least $US 3 trillion per year, or 11 per cent of the annual world economic output (www.usaid.gov/our_work/environment/biodiversity).

This integration of conservation and development found a significant funding source in the US Agency for International Development (USAID), which expanded into the area in the 1980s by mandate from Congress. In addition to sponsoring individual projects and conferences in lower income countries, USAID administers a sizeable peer-reviewed research grant programme. The centrepiece of the latter is the Conservation of Biological Diversity project. This has two main components: cooperative funding of National Science Foundation (NSF) grants for research that contributes to the conservation of biodiversity in developing countries, and core funding for the Biodiversity Support Program, a consortium formed by the World Wildlife Fund, The Nature Conservancy and the World Resources Institute. USAID projects have been carried out in Latin America, the Caribbean, Africa and North Africa as well as Europe and Asia (Alpert 1993: 630). By the early 1990s, the agency was investing about $100 million a year in biodiversity programmes around the world (Angier 1994). Today, it supports conservation activities in 64 countries.

The assembly of biodiversity loss as an environmental problem benefited greatly from the participation of several well-known scientific news entrepreneurs or champions who were extremely active in promoting it both within and beyond the parameters of science.

The Ehrlichs, Paul and Anne, had already achieved a measure of fame (as well as notoriety) in the late 1960s and early 1970s for their campaign to make overpopulation the centrepiece of the environmental crisis. Subsequently, the two biologists turned their attention to the problem of biodiversity loss. In 1986, they founded the Center for Conservation Biology at Stanford University with Paul Ehrlich as president (see above). In 1981, the Ehrlichs published *Extinction*, one of several high-profile books that appeared on the topic of endangered species and biodiversity around this time. Here they infused the biodiversity problem with a moral dimension using the 'Noah Principle' to claim that the foremost argument for the preservation of all non-human species is the religious belief 'that our fellow passengers on Spaceship Earth...*have a right to exist'*.

A second major champion of the conservation of biodiversity was the celebrated Harvard entomologist Edward Wilson. Widely known as one of the founders of the field of 'sociobiology', Wilson is also a 1978 Pulitzer Prize-winning author whose bestselling book, *The Diversity of Life*, was carried as a selection by book clubs across the US and Canada. In his autobiography, Wilson reports he was 'tipped into active engagement by the example of my friend, Peter Raven', who had been writing, lecturing and debating the issue of mass extinction since the late 1970s. Among his contributions, Wilson edited a key collection of articles arising out of the 1986 National Forum on Biodiversity under the title *Biodiversity* (1988); this became one of the bestselling books in the history of the National Academy Press (Wilson 1994: 358).

Other key figures in assembling the problem of biodiversity loss were Raven, director of the Missouri Botanical Garden, Norman Myers, well-known international conservationist who published the book *The Sinking Ark* in 1979, and Michael Soulé, a founder and populariser of the discipline of conservation biology.

Longtime friends who had similar interests, moved in the same circles and often did fieldwork in the same areas (Mazur and Lee 1993: 703), Ehrlich (Paul), Raven and Wilson were involved in many of the same endeavours to promote biodiversity loss as a global environmental problem. Ehrlich and Raven organised and chaired panels at the 1986 forum in Washington. In 1989, Raven and Wilson gave expert testimony before the US Senate Subcommittee on Environmental Pollution. Wilson and Ehrlich were contributors to the special biodiversity issue of *Science* in August 1991. And all three scientists were founding members of the Club of Earth, an activist group formed to bring scientific attention more quickly to important but neglected environmental problems (Brown 1986). Mazur and Lee observe of this trio:

> Their research productivity, their eminence and their social and institutional contacts gave them strong voices within the scientific establishment and good access to Federal and private sources of funding, which supported both their scientific and policy efforts. (1993: 703)

Wilson (1994: 357–8) refers to a 'loose confederation of senior biologists that I jokingly call the "rainforest mafia"' whose members besides Raven, and himself, included Jared Diamond (who a decade later went on to become the bestselling author of several influential books on societal collapse), Thomas Eisner, Daniel Jantzen, Thomas Lovejoy and Norman Myers and who were instrumental in advancing claims about the importance of biodiversity loss.

Presenting the claim

In presenting biodiversity as an environmental claim and keeping it on the public agenda, proponents face three formidable problems (McNeely 1992: 25).

First, unlike some other environmental problems such as toxic dumps or oil spills at sea, there is no easily identifiable opponent against which public opinion can be galvanised. Instead, the root causes of biodiversity loss are found in basic economic, demographic and political trends including the relentless human demand for commodities from the tropics, runaway population growth and the escalating debt burdens of Third World nations (McNeely *et al.* 1990b)

Second, the loss of biodiversity has no immediate impact on human lifestyles in the First World nations where the resources that could be applied to acting upon the problem are concentrated. Indeed, with the exception of a small number of 'charismatic megafauna' – whales, gorillas, whooping cranes, bald eagles – most threatened organisms consist of creatures such as fungi, insects and bacteria that most people would not hesitate to step on (Mann and Plummer 1992: 49). This problem is even more exaggerated at the system level where, as Noss (1990) has sardonically observed, 'you can't hug a "biogeochemical" cycle'. As Mazur and Lee note:

> The plights of these [charismatic] animals became salient through popular books, television documentaries such as those produced by the National Geographic Society, and news coverage of a few effective spokespersons including Jacques Cousteau, Brigitte Bardot, Roger Tory Peterson and Jane Goodall, who usually specialized in a single type of animal.
>
> (1993: 701)

Third, the collective benefits of taking action are notably imprecise. At best, conservationists can speculate that somewhere in the vanishing rainforest lies the cure for cancer or AIDS but there are no ironclad guarantees. By contrast, the costs of implementation are more apparent and onerous especially on the domestic front in developed nations. As a result, public attention often begins to lag when the visible costs seem to outweigh the immediate benefits. Public support can be further eroded by profile controversies such as that which occurred in the United States in the late 1980s and early 1990s over the fate of the Northern spotted owl.[4]

Claims-makers have addressed these difficulties by adopting a 'rhetoric of loss' (Ibarra and Kitsuse 1993). Public statements by conservation biologists and other policy entrepreneurs stress that we are 'at a crossroads in the history of human civilization' (McNeely *et al.* 1990b: 40). Failure to act decisively is equated with turning down the road to chaos or driving a business into liquidation. Many of these metaphors are borrowed from the rhetoric of another environmental problem – overpopulation. Once again, we are depicted as rapidly approaching the 'limits of growth', thereby running the risk of surpassing the 'carrying capacity' of the planet. Lester Brown, president of the World Watch Institute, a well-known environmental think-tank, uses the rhetorical motif of a 'race' to describe how the momentum inherent in population growth with its attendant problems for biodiversity is pushing us rapidly towards a catastrophic finish line (1986).

A sample of rhetorical statements on biodiversity loss by prominent environmentalists/scientists

'We have not inherited the Earth from our parents, we have borrowed it from our children.'

Peter Raven in Congressional testimony (see Kellert 1986) and IUCN *World Conservation Strategy*, Introduction (1980)

'We are in a race. Maybe we should call it a contest.'

Lester Brown (1986)

'We're treating the world as a business in the process of liquidation.'

Peter Raven, cited in Gooderham (1994)

'The future well being of human civilization and that of many of the 10 million other species that share this planet hangs in the balance.'

McNeely *et al.* (1990b)

'In the last twenty-five years or so, the disparity between the rate of loss and the rate of replacement [of species and populations] has become alarming; in the next twenty-five years, unless something is done, it promises to become catastrophic for humanity.'

Ehrlich and Ehrlich (1981)

'Elimination of lots of lousy little species regularly causes big consequences for humans, just as does randomly knocking out many of the lousy little rivets holding together an airplane.'

Jared Diamond (2005)

The rhetoric of 'catastrophe' was further enhanced through linking it to the fate of the dinosaurs.[5] In 1980, two eminent scientists from the University of California at Berkeley, Nobel prize-winning physicist Luis Alvarez and his geologist son Walter, proposed that the dinosaurs had perished as the result of climate changes brought about by an asteroid which had crashed on earth sixty-five million years ago. Few scientific theories have attracted more public interest as quickly as did this controversial claim, a fact not lost on biodiversity activists who often used the dinosaurs as a point of comparison (Mazur and Lee 1993: 703). Similarly, a television advertisement sponsored by the Humane Society of Canada in the mid-1990s, proclaimed: 'it is the greatest extinction rate since the end of the dinosaurs'.

A subsidiary idiom here is that of 'entitlement'. Thus both Raven, in his testimony before the 1981 Congressional committee and the IUCN in the introduction to *World Conservation Strategy* reiterate a memorable slogan to the effect that 'we have not inherited the Earth from our parents, we have borrowed it from our children'.

Running parallel to this 'doomsday' rhetoric is a second type of claims language that stresses the positive economic benefits of preserving diverse habitats. Using warrants that are loaded with financial figures, proponents favour a 'rhetoric of rationality' (Best 1987).

For example, just over a decade ago, Walter Reid, a vice-president of the World Resources Institute, wrote this about the 'economic realities of biodiversity':

> Currently some 25 per cent of US prescriptions are filled with drugs whose active ingredients are extracted or derived from plants. Sales of these plant-based drugs amounted to $4.5 billion in 1980 and an estimated $15.5 billion in 1990...In Europe, Japan, Australia, Canada and the United States, the market value for prescriptions and over-the-counter drugs based on plants was estimated to be $43 billion in 1985.
>
> (1993–4: 49)

Significantly, this rhetoric uses the language of frontier development, for example, referring to 'bioprospecting' (Eisner 1989–90; Reid *et al.* 1993) or 'biotic exploration (Eisner and Beiring 1994). It is suggested that somewhere in the 'biotic wilderness' scientists will find an equivalent of Madagascar's rosy periwinkle with its famous cancer-fighting properties (Eldredge 1992–3: 92)

This depiction of tropical rainforests as the cradle of tomorrow's pharmaceutical medicine has recently spread into the arena of popular culture. In the American motion picture *Medicine Man*, Sean Connery played a maverick scientist who discovered a miracle cure for cancer among the canopies of the South American rainforest, only to have his research site flattened by the bulldozers of a road-building crew. And in *Day of Reckoning*, a 1994 action movie with a 'new age' flavour, an adventurer hunts for a rare plant with medicinal powers in the rainforests of Burma. As W.H. Hudson's romantic novel *Green Mansions* illustrated nearly a century ago, the human threat to the diversity of life in tropical ecosystems can make a compelling drama, with strong moral and spiritual overtones.

Contesting the claim

While individual countries undertook unilaterally to protect endangered species and habitats, it became obvious at least 35 years ago that concerted global action on biodiversity loss required some type of coordinated multilateral agreement. In fact, an International Convention on Biological Diversity was first proposed in 1974 and active planning for such an accord began in earnest in 1983 (Tolba and El-Kholy 1992). This process culminated in 1992 with the preparation of a Global Biodiversity Strategy under the auspices of three agencies: the United Nations Environment Programme (UNEP), the World Conservation Union (IUCN) and the World Resources Institute (WRI). In order to carry out the recommendations of this strategy, it was proposed that a Convention on Biological Diversity be put forward at the United Nations Conference on Environment and Development (UNCED) in Rio in June, 1992. By all accounts, this convention was conceived in a medium of considerable controversy, especially with regard to the question of access to genetic resources in Southern hemisphere nations.

At the core of the treaty could be found a basic tension between two conflicting commitments. On the one hand, the proposers wished to provide a mechanism whereby the international conservationist community could directly intervene in situations where sensitive environmental areas with diverse biological resources were threatened, notably in the tropical rainforests of Brazil. On the other hand, target nations were not eager to lose their national autonomy and give up the right to make their own decisions, particularly with regard to development projects. As compensation for allowing outsiders to infringe their traditional national sovereignty, less developed nations wanted something in return, specifically, financial resources and the transfer of technology from the industrial nations of the North.

Furthermore, the Southern nations wanted to use the occasion to tighten up access to their genetic resources that theretofore had been more or less free to all comers. According to Article 15 of the Convention, nations were to have sovereign rights over their genetic resources and grant access only on mutually agreed terms and with 'prior informed consent'. Other provisions attempted to deal with some of the more potentially exploitative aspects relating to the appropriation of Third World genetic resources by multinational biotechnology companies. Increasingly, these firms have begun to prospect tropical habitats for unusual species of plants and animals, to 'borrow' key genetic material, bioengineer and patent it and then license the improved product back to the country of origin at a hefty profit. Accordingly, the South argued for access to the results and benefits of biotechnologies developed in connection with those genetic materials that have been exported, specifically in the form of continuing royalties and technology sharing.

Even at the pre-summit stage, a number of these points were contested. For example, an earlier draft of the convention had called for two global lists – a Global List for Biological Diversity and a Global List of Species Threatened with Extinction on [a] Global Level – that would have spelled out in priority fashion the commitments that were required of signatories to the treaty. However, during the final negotiations at Nairobi, Kenya, leading up to the Rio Summit, these references to global lists were removed, a measure that was strongly contested by many delegates including the leader of the French delegation who refused to sign the final act. Similarly, a provision that would have furnished free 'scientific access' to genetic resources in biologically diverse nations was dropped from the final convention (Barton 1992).

At the summit itself, the United States incurred the wrath of other participants by refusing to sign the Biodiversity Convention, even though 153 other countries did so and the Secretary of the Environment himself was in favour. This appeared to be the result of considerable pressure on President Bush from American biotechnology trade associations, which objected to the provisions that would have meant that US firms must pay continuing royalties and share new patents and technological secrets with nations whose biological resources are the source of new products (Susskind 1994: 182).

The Biodiversity Convention was challenged by a third party that was not present at either the negotiations or the Summit. This was a coalition of farmers, ecological activists and others from Third World nations who felt that local people had been excluded from the formulation of the treaty, especially the provisions relating to intellectual property rights. Their absence has subsequently been noted annually in the *Report of the Global Biodiversity Forum* 'Background' section with the statement: 'However, the process prior to and following the development of the CBD [Convention on Biological Diversity] did not in general allow for the full participation of all those interested and affected'.

The best-known spokesperson for this movement is the Indian eco-activist Vandana Shiva. Shiva and her movement have attempted to wrest 'ownership' of the problem of biodiversity loss from conservation biologists, non-governmental global environmental organisations and government negotiators who they accuse of assuming a mantle of leadership that is not theirs to wear. In particular, they object to the exclusion of the original donors of genetic resources – Third World farmers – from the exchange of resources and knowledge which the Convention governs. The basic problem, Shiva states, is that:

> those 'selling' prospecting rights never had the rights to biodiversity in the first place and those whose rights are being sold and alienated through the transaction have not been consulted or given a chance to participate.
>
> (1993: 559)

Shiva observes that even in the case of the 1991 agreement between Merck Pharmaceuticals and INBIQ, the National Biodiversity Institute of Costa Rica, a much-heralded and publicised example of how it is possible for multinational corporations to compensate the Third World for its genetic resources, the people living in or near the national parks in Costa Rica were not consulted, nor were they guaranteed any economic benefits. Rather, the agreement was forged between Merck and a conservation group formed at the initiative of a leading American conservation biologist Dan Janzen, who, it will be recalled, was a member of Wilson's 'rainforest mafia' (Shiva 1993: 559).

Opponents of 'commercialised conservation' (Shiva 1990: 44) have proposed the formulation of an alternative form of intellectual property, the *Samuhik Cyan Sanad* or Collective Intellectual Property Rights (CIPRs). These collective patents invest the right to benefit commercially from traditional knowledge in the community that developed it. Furthermore, it is demanded that multinational companies seeking to utilise Third World genetic resources be compelled to deal through the village organisations who would hold title to these CIPRs. Failure to do so, it is claimed, would constitute 'intellectual piracy' (Shiva and Holla-Bhar 1993: 227).

Shiva's challenge has not gone unnoticed. At the 7th Session of the Global Diversity Forum, held in Harare, Zimbabwe in June 1997, the official 'Statement' prepared by Forum participants included the following paragraph:

> Participants recommended that CITES mechanisms be developed to incorporate local and traditional knowledge and local participation in decision-making at all levels including in the national scientific bodies and international forums. National governments should be encouraged to involve local communities in the development and implementation of CBD and CITES strategies. All parties should be encouraged to include assessments of potential impacts on local communities when proposing changes to existing Conventions. Improvements are needed in the national and international processes for carrying out the goals of the Conventions to reflect the rights and aspirations of local communities.
>
> (Global Biodiversity Forum 1997)

Most recently, international biodiversity discourse has shifted towards the reconciliation of conservation and poverty reduction through development, two goals that have often been depicted in the past as being at odds with one another and driven by different moral agendas. Thus, the IUCN's director general now describes protected areas as 'islands of biodiversity in an ocean of sustainable development' (cited in Adams *et al.* 2004: 1146). And, the Millennium Development Goals (MDGs), adopted at the 2000 UN Millennium Summit, links environmental sustainability with poverty eradication, education, gender equality, reduced child mortality and the creation of a global partnership for development. Conservation and the sustainable use of biodiversity are becoming increasingly perceived as critical to the full achievement of the MDG goals (Timmer and Juma 2005: 27).

One high-profile programme that has attempted to convert this rhetoric into achievable gains is the *Equator Initiative*, launched in 2002. The initiative is a partnership among local grassroots groups in countries along the equatorial belt, the United Nations and the UN's global development network. Its centrepiece is the Equator Prize, awarded to local community partnerships that work simultaneously toward sustainable income generation and environmental conservation. Some past prize recipients include the Green Life Association of Amazonia (AVIVE) in Brazil which focuses on the sustainable extraction and marketing of medicinal and aromatic plant species; the Genetic Resource, Energy, Ecology and Nutrition (GREEN) Foundation in India which works through a network of women's farming groups called *sanghas* to improve food security by conserving indigenous seeds and establishing community seed banks and home gardens; and the Suledo Forest Community in Tanzania that harnesses local knowledge of the forests to regulate poaching and promote sustainable silviculture (see Timmer and Juna 2005).

Despite the promise of such prizewinning projects, the dual goals of biodiversity loss and poverty eradication are not always fully compatible. Projects that seek to integrate conservation and development have tended to be 'overambitious and underachieving' and lasting positive outcomes remain 'elusive' (Adams *et al.* 2004: 1147). For example, ecotourism ventures, one popular type of project undertaken by Equator Prize winners, are risky insomuch as they are vulnerable to international tourist fads; create pressure to build hotels and leisure facilities that negatively impact the resource base on which the community depends; and may fail when other local communities choose ecotourism as their source of alternative livelihood, thereby saturating the market (Timmer and Juma 2005: 31–2). Other projects falter when they are hijacked by local élites as a way of solidifying their interests. As Timmer and Juna (2005: 35) note, 'ignoring differences in values, perspective and power within a community and [the] differential access that community members have to layers of political decisionmakers leads to inaccurate assumptions about the ease by which collective decisions at the local level can be made'.

Conclusion

The rapid ascent of biodiversity loss in the international arena is somewhat surprising. While extensive, research on biodiversity largely navigates uncharted waters. Of the 1.4 million species known around the time that the Convention on Biological Diversity was adopted (this has since risen to approximately 1.75 million (Spray and McGlothlin 2003a: xvi), only five per cent can be considered 'well known' and the relationships between many of them are a mystery (Gooderham 1994: A–12). The theory that underlies ecosystem diversity is based primarily on small-scale studies of ponds projected on the larger screen of nature. The benefits of acting boldly are not precisely documented. The costs are considerable, not only financially, but also in behavioural terms. If large-scale biodiversity protection is to be implemented, the number and range of people affected and the extent of change required are considerable (Balch and Press 2003: 124–5). The ownership of the problem is disputed with multiple claimants.

Yet despite these drawbacks biodiversity became a major environmental theme in the 1990s. There are several factors that account for this.

First, it is not purely an environmental problem but is simultaneously an economic and political question. For business, biodiversity has the potential to be made into a valuable resource that can generate a tidy profit. For Third World governments, it is both a source of foreign exchange and a window through which First World biotechnology can be accessed. For small farmers in India and other poor nations, it is a means of empowerment and resistance to the creeping power of global capital (Shiva and Holla-Bhar 1993).

Second, biodiversity loss constitutes a socially constructed environmental problem that has brought together two well-established organisational sectors: the international development establishment and the global conservation network. Nested in a web of NGOs, notably those connected to the United Nations, it has an institutional momentum extending beyond that which is able to be generated by single environmental movement organisations such as Greenpeace and Friends of the Earth which have more of an 'outsider' status.

Third, the biodiversity problem has not been constructed from scratch but has flowed out of the already long-standing problem of endangered species. The two problems are to a large extent symbiotic and synergistic. Biological diversity gives species endangerment and extinction a theoretical grounding that it previously lacked. The example of endangered species provides biological diversity with a specific focus and an emotional resonance that the more general issue often lacks. Furthermore, the preservation of diversity furnishes a rationale for action in rancorous environmental disputes such as those that have raged in recent decades over Great Whale River and the Clayoquot Forest in the Canadian North and West (Suzuki 1994b)

Finally, the location of biological diversity at the centre of the discipline of conservation biology means that it has been buffered against the 'issue attention cycle' (Downs 1972) that affects a great many other environmental issues. Furthermore 'the biodiversity debate has not been embroiled in the kind of scientific disputes that have occurred in debates on acid rain, the depletion of the ozone layer and climate change' (Valiverronen 1999: 407). Conservation biology provides biodiversity loss with a centre of gravity around which it can revolve, rotating out into the realm of international diplomacy and conflict but stabilised by the continual pull of research within this speciality area.

10 Towards an 'emergence' model of environment and society

A twilight zone

On Boxing Day, 26 December 2004, a 'monster' tsunami slammed into coastal regions across the Indian Ocean bringing an almost unprecedented level of death and destruction. Triggered by an earthquake off the coast of Sumatra, the tsunami cut a wide swath impacting seventeen countries, most notably Indonesia, Thailand, Sri Lanka, India, Malaysia, Burma, the Maldives archipelago, Andaman and Nicobar Islands and the western coast of Africa. Even today, the final, official death toll is under constant revision, with recent estimates putting it in the order of 176,000 deaths. The highest loss of life was in the Indonesian district of Aceh on the northern tip of Sumatra. Meulaboh, the town nearest to the quake epicentre, was totally devastated – 80 per cent of its buildings were destroyed.

While those in the affected areas were no doubt too traumatised to engage in much reflection, academic and media commentators struggled to define the nature and meaning of the event. Was it a 'natural' environmental catastrophe on a massive scale or did it have some 'human' cause? A correspondent for the *Financial Times* stated quite plainly that 'the Indian Ocean tsunamis were caused by an underwater earthquake, and had nothing to do with global warming and climate change'; but, immediately qualified this by adding 'however, they may give a foretaste of some of the disasters that experts are predicting as a result of climate change' (Harvey 2004). Eco-activist Vandana Shiva (2005: 22–3) had a less nuanced view, warning readers of *The Ecologist* that the lesson from the tsunami was that this is a foretaste of what rising sea levels will look like if 'the rich North cannot afford to take action to reduce CO_2 emissions and work towards reducing the impact of climate change'.

Others noted that a contributing factor might have been overdevelopment, especially in the coastal tourist zones of Thailand. The International Tsunami Survey Team in Sri Lanka reported a number of instances where human development likely magnified the ease with which the tsunami penetrated ('ranup') inland. For example, one resort that had previously removed some of the sand dune seaward of its hotel suffered far greater damage (including destruction of the hotel) compared to neighbouring areas located behind unaltered dunes. And the *Sumudra Devi*, a passenger train, was derailed and overturned by the tsunami wave, killing more than 1,000 passengers in an area where substantial coral mining had occurred related to tourist development (Liu *et al.* 2005). In similar fashion, 900 fewer people were reported dead or missing in the Maldives, which regulates coral reef management, than in Phuket, Thailand, where coastline coral and mangroves had been replaced by aquaculture and hotels ('Tsunami's impacts' 2005).

In the hours, days and months after impact, the tsunami generated a host of medical and social problems, some of which were beyond previous experience. Largely spared the water-borne illnesses (hepatitis, typhoid, malaria) that health authorities feared would follow, thousands appear to have been stricken with 'tsunami lung', a disease caused by a mixture of bacteria in the saltwater and mud that the tsunami churned up and which people caught in the waves swallowed (Zamiska 2005). Oxfam reported that in the Indian state of Tamil Nadu the tsunami disproportionately took the lives of females and children (the males were out to sea on their fishing boats). This has drastically changed the demographic makeup of these fishing villages, thereby altering the social composition and responsibilities within individual families. One unexpected result here is that local authorities have announced a reversal of a government sponsored sterilisation programme, aimed at cutting population growth; surgery to reconnect a woman's Fallopian tubes will now be paid for by the state.

In Northern Sumatra, many residents have lost both their land and their personal identity. This has necessitated the formulation of new procedures. Thus, to reclaim your identity:

> First you need to find two people to whom you are not related who can vouch for you. Then you need to see the village or neighbourhood chief. Then you need to take his letter to the sub-district chief. Only then do you get an identity card.
>
> (Aglionby 2005)

As is typical in disaster situations, a whole range of volunteer initiatives and organisations have sprung up. Some are attached to existing NGOs such as Oxfam, Save the Children and the Red Cross, while others such as the Khao Luk Volunteer Centre in Thailand arose specifically in response to the situation.

It is common to observe the appearance of an 'altruistic community' in the first days of the emergency where existing conflicts and animosities are put aside and people reach out to help others in a spirit of cooperative reconstruction. In other cases, however, notably in technological disasters, we see the emergence of a *corrosive community* (Freudenburg and Jones 1991). Here, people are set against one another rather than bind together in a sense of common struggle and recovery; relations become caustic; and conflict predominates (Clarke 2003: 132). Even if this does not occur, pre-existing fault lines of ethnicity, class, race and gender often re-emerge after an initial grace period. At least two of the areas impacted by the tsunami, North Sumatra and northeast Sri Lanka, have been sites of rebellion or civil war or both. In the immediate aftermath of the tragedy, some evidence of a truce was observed, but this is already beginning to erode. In Sri Lanka, for example, there have been disputes over distributing the nearly $US3 billion in promised foreign aid, culminating most recently in a Supreme Court-ordered freeze of the aid distribution agreement between the Government and the rebel Tamil Tigers (Goodspeed 2005).[1]

In the resort areas of Thailand, the imperatives of global tourism have created an especially uncertain situation. Recalling what he observed during the first days of the disaster, CBC television reporter Sasa Petracic noted that there was a dual response based on the identity of the victims. Tourist corpses were immediately refrigerated and a team of forensic technicians with laptop computers sprang into action, identifying the deceased and communicating with the next of kin abroad. By contrast, local Thai victims were collectively buried under a giant earth pile, later to be disinterred. Petracic remembered that the first business to be rebuilt at one of these resort towns was Starbucks. What this indicated was an awareness on the part of the Thai authorities that the international media coverage was concentrated, at least in part, on the fate of vacationers from Western nations.

On a visit to Khao Lak on Thailand's resort coast six months after the tsunami, another journalist, John Bussey (2005) observed a kind of twilight zone where the search for human remains continues, mourning is still in the early stages, and the economy is moribund. Some spas and luxury hotels have been rebuilt for foreign tourists who may not return in significant numbers. 'Sometimes it isn't clear', Bussey says, 'whether the region is rebuilding as a resort, or, for the time being, a memorial'.

How might the Indian Ocean tsunami and its effects most helpfully be conceptualised by environmental sociologists?

To begin, let us turn to Raymond Murphy's (2004) seminal treatment of another disaster, the ice storm that impacted parts of Quebec and Eastern Ontario in January 1998. Murphy uses the metaphor of a 'dance' to describe the interactive relationship between nature and society. Sometimes, nature takes the lead and humans react and improvise after nature's moves in this dance. Other times, humans take the lead and choreograph a response in anticipation of nature's moves. In the case of the ice storm, nature issued an extreme 'prompt' that was, at least initially, ignored or denied. To urban residents wholly dependent on a connection to the North American power grid, this was a catastrophe since their heat, light, electricity and even drinking water was dependent on the technology. By contrast, for the small, decentralised Amish communities in northern New York State who used wood stoves for their cooking and heating needs and milked cows by hand, the ice storm caused minimal problems. Murphy concludes that the ice storm disaster 'resulted not from freezing rain per se, but rather from the vulnerability of the infrastructure that modern society had constructed and upon which it had become dependent' (p. 257).

The tsunami case calls up some of the same points. For example, in the Andaman Islands 'stone age' tribal groups were unhurt, having fled the coast before the disaster. In the absence of a technologically sophisticated tsunami warning system, they were evidently alerted by the unusual flight behaviour of wild and domesticated animals that were observed to act in fearful, anxious or unusual ways days or hours before the onset of the wave (Sheldrake 2005). Furthermore, as has been noted, the extent of the damage and loss of human life was inflated by an ill-considered set of decisions in recent years to alter the natural ecology of the sand dunes, mangroves and coral reefs, all in the name of 'tourist development'. In such instances, nature's 'prompt' can and should inspire a process of environmental learning.

Perhaps because he is preoccupied with fleshing out the basic framework of a new 'constructionist realist' approach to environmental sociology, Murphy does not extend these important notions of prompts, improvisation and creative movement too much beyond the nature/society nexus. Yet, they have considerable applicability within a wide spectrum of environmental events, arenas and policy zones, including but by no means restricted to disaster episodes.

To capture these dynamics more fully, I am proposing an approach to environment and society that pivots on the concept of *emergence*.

Emergence denotes process, flow, adaptation and learning. In the physical and biological sciences, it is associated with what has come to be known as 'complexity theory'. In his book, *Emergence*, American cultural and technology commentator Steven Johnson (2001) says that emergence is what happens when an interconnected system of relatively simple elements self-organises to form a more intelligent, more adaptive higher-level behaviour. He illustrates this using the disparate cases of ant colonies, human immune systems, media events and urban neighbourhoods. Within the last decade, Johnson maintains, emergent complexity has entered a new phase in which self-organising systems are becoming the state of the art in software applications, video games and even music.

One recent social science adaptation of complexity theory can be found in the politics of international relations. Drawing on 'the science of complex systems', Harrison and Bryner (2004: 343–4) sketch out a theory of 'emergence processes' as applied to the production of international environmental policy. Complex systems, they explain, are emergent, dynamic and potentially nonlinear (disproportionately sensitive to small changes in internal and external conditions). This, they say, has several important implications.

First, international environmental policy should be seen as being not only the creation of states, but, rather, the product of a complex interaction of many related processes 'including the negotiated conclusions of authoritative scientific reports, international discourse between states, the emergent demands of interest groups and the public through domestic political processes, and the beliefs and preferences of governments and leaders' (p. 344). Second, scientific evidence can be a primary tool of persuasion and even individual scientists can make a difference. Third, social learning is possible, since international environmental policy is 'an emergent property of complex and dynamic processes' (p. 345).

What all of these ideas of emergence have in common is a realisation that social organisation and the production of knowledge are fundamentally fluid, dynamic, and adaptive. There is also a strong suggestion that they percolate from the grassroots rather than pass from the top downwards. The case of the Indian Ocean tsunami offers an excellent opportunity to study emergence in action. With many existing certainties washed away, new actions and formations are possible, and in some cases, even necessary, at least for a while.

While conceptualised rather differently, emergence has a long, if not widely known, history in several areas of sociology. Although it never quite achieved paradigmatic status, emergence theory has been around for nearly half a century in the sociology of collective behaviour and social movements. In recent years, it has been recast as a significant but usually unacknowledged presence in cultural approaches to social movements, notably those relating to collective identity formation and social learning. From a policy perspective, emergence theory has been most influential in the sociology of disasters, most recently in relation to institutional and grassroots responses to the World Trade Center attacks.

Sociological foundations of emergence

Emergent norm theory and collective behaviour

Emergence theory, or as it was initially known *emergent norm theory*, was introduced in 1957 in the first edition of Ralph Turner and Lewis Killian's foundational text *Collective Behavior*. Borrowing concepts from the small-group studies of Asch (1951), Lewin (1947) and Sherif (1936), Turner and Killian proposed that a member of a crowd acts in a particular manner not because of a blind propensity to imitate, nor as the result of being 'infected' by a 'contagion', but rather because a certain course of action is perceived as being appropriate and required (Milgram and Toch 1969: 553). Group pressures toward conformity thus operate in collective behaviour situations as well as in everyday settings. What is characteristically unique about a collective behaviour episode is that the situation is ambiguous or undefined, and therefore existing norms fail to provide significant guidance. As a result, the crowd or other collectivity is forced to innovate, together forging its own guidelines for behaviour. Frequently, one or more innovators ('keynoters') suggest a course of action and a consensus develops that this be considered as appropriate. In ambiguous situations, for example, the aftermath of a disaster or during a civil disturbance, reliable information is difficult to obtain and collective actors therefore often rely on rumours to supply the appropriate cues. Shibutani (1966), in a memorable turn of the phrase, labelled rumours as 'improvised news'.

Emergent norm theory was undoubtedly an improvement on existing theories that all more or less incorrectly depicted collective behaviour as a non-rational or irrational response that 'precludes any examination of innovations or learning on the part of collective actors' (Cohen 1985: 672). Indeed, behaviour such as rumour communication was depicted in emergent norm theory as constituting a rational search mechanism. Nonetheless, there were a number of conceptual difficulties. Emergent norm theory said little about the *content* of norms that arise in crowd situations and failed to account for the primacy of one norm over another (Milgram and Toch 1969: 555). Furthermore, Turner and Killian's approach tended to be tautological, that is, emergent norms constituted both the definition of, *and* the explanation for, collective behaviour (Tierney 1977: 14).

Several decades later, Turner and Killian revisited emergent norm theory, clarifying their original views. Killian (1980: 284) allowed that 'perhaps the choice of the term [emergent norm] has been somewhat misleading as it was too narrow'. Rather than being just a prescription for or prohibition against a specific action, Killian claimed that he and Turner had really meant the term emergent norm to include an extensive complex of factors: rules applicable to a situation; explorations of the situation; evaluations of potential actors; and a shared conviction of right which constitutes a norm, sanctions behaviour consistent with the norm, inhibits behaviour contrary to it, justifies proselytising, and requires restraining action against those who dissent (1980: 284).

Following on this, in the third revised edition of *Collective Behavior* Turner and Killian (1987: 33) reiterate that 'the concept of norm as used here does not refer merely to a rule or a precise behavioral expectation; rather, it encompasses a complex of factors, including indications of the salient features of the situation and typifications of the actors presumed to be involved'. This re-conceptualisation expanded the potential applicability of the perspective to a much wider spectrum of collective action. Despite this, Tierney (1977: 16) criticised this revised version as constituting a 'virtual catchall under which all types of collective ideational phenomena...cognitions, expectations, beliefs, symbols, definitions of the situation and ideological systems are subsumed'.

Emergence theory and disaster research

In the 1970s, an even broader conceptualisation of emergence developed, most notably in the work of Russell Dynes and E.L. Quarantelli and their students at the Ohio State University Disaster Research Center (DRC) on community disasters and collective behaviour.

Brouillette and Quarantelli (1971) distinguished between 'emergent structures' and 'emergent tasks' during emergency situations. Two years later, in an article in the *American Journal of Sociology*, Weller and Quarantelli (1973) overlay a variable that relates to groups (whether the collectivity involved consists of enduring or emergent social relationships) onto the variable that had traditionally been used to define collective behaviour – whether the norms guiding behaviour are institutionalised (enduring) or emergent. Collective behaviour was thus depicted as taking the form of emergent norms, emergent social relationships or both. While praising this typological scheme as 'the most successful conceptualization [of collective vs. non-collective behaviour] thus far', Gary Marx (1980: 267) also expressed some reservations: (1) the typology mixes elements of structure and behaviour, and classification and explanation; (2) it is difficult to examine causal relations between the categories because one can influence the other (for example, enduring or emergent social relations are likely to condition the type of norms that are present).

In his seminal book *Organized Behavior in Disasters*, Russell Dynes (1970) noted that in community disasters such as floods, tornadoes and hurricanes, the normatively guided response can include both emergent patterns of authority and emergent lines of communication. He identified four distinct types of organisations based on whether any changes in their structure and/or tasks were evident during and after the disaster period. *Established* disaster-related organisations, for example community fire departments, basically maintain the same tasks and structures during a disaster as they do under normal conditions. *Expanding* organisations, for example the Red Cross, carry out the same tasks but become larger, necessitating a transformed structure. *Extending* organisations, for example, construction companies, keep more or less the same structure but engage in different tasks. Finally, *emergent* organisations arise in response to the experience of the disaster situation. They are characterised both by emergent tasks and emergent structure.

Subsequently, in a special 1973 issue of the *American Behavioral Scientist* (*ABS*) devoted to organisational change and group emergence during the urban civil disturbances (riots, campus protests) of the 1960s, DRC researchers dealt with several of these emergent organisations. Ponting (1973) discussed 'Rumour Control Centers' (RCCs) most of which were set up in 1968 in order to alleviate potential social unrest caused by the spread of rumours in ghetto neighbourhoods. Most of the RCCs studied by Ponting were 'intermittent organisations' (Etzioni 1961: 288–96), that is, they remained dormant or at a low level of activation until tensions began to spike in the community. Teuber (1973) looked at the emergence of 'Human Relations Commissions' (HRCs) established by big city mayors to 'cool off' conflicts among racial and cultural groups and reduce underlying strains. Forrest (1973) presents a case study of an 'Interfaith Emergence Center' (IEC) in Detroit, Michigan that began as a crisis telephone hot line but quickly transformed into a broker connecting individual needs with community resources and services.

Finally, Dennis Wenger, who had contributed an article to the *ABS* issue on the structural changes that occur in police departments during civil disturbances, expanded the repertoire of emergent phenomena even further. In examining community structural adaptations in a disaster setting, Wenger (1978) identified four emergent forms: emergent values and beliefs, emergent normative structures, emergent organisational structures, and emergent power structures. This latter typology is especially useful in constructing an emergence approach to environmental sociology.

More recently, Tierney has re-introduced the notion of emergence in relation to the organisational response to the events of 11 September 2001. Commenting on the official recommendations of the *9/11 Commission Report*, she is critical of the authors' conclusion that the responding emergency agencies first on the scene at the World Trade Center failed to exert 'command' and coordinate the response. The widely subscribed-to notion of disaster command is inconsistent, Tierney argues, with what disasters typically are:

> complex occasions characterized by a high degree of ambiguity often coupled with extreme urgency, that require extensive improvisation and that call for more autonomy, rather than less on the part of organisational entities involved in the response. (2004: 117)

As indicated in the earlier DRC research, Tierney (2003: 40–2) observed that group emergence occurred in a variety of contexts. In some instances, groups carried out their activities entirely outside the formal response structure, as when local residents prepared and delivered meals to emergency workers at Ground Zero. In another common pattern, blended networks formed consisting of existing agencies and volunteers. For example, safety inspections in and around the impact site were carried out by city building and safety officials, working alongside volunteer structural engineers. Improvisation was the order of the day. In the previous example, the volunteer engineers introduced an adapted version of a rapid damage-screening protocol that had originally been developed by engineers in California for conducting building safety assessments following earthquakes. In another episode reminiscent of the Dunkirk evacuation during the Second World War, several hundred thousand people were evacuated from Lower Manhattan by water via an emergent network of private and publicly owned watercraft.

Tierney stresses that the response to the September 11 tragedy was effective because it was flexible, adaptive and locally-based, rather than centrally directed and controlled.

By contrast, initial media reports concerning the four bomb blasts that impacted London's transit system on 7 July 2005 suggest that rescue and evacuation activities were more dependent on direction from police, fire, ambulance and London Underground crews in orange, hazardous materials suits who were following a pre-established emergency plan. Nevertheless, some emergent and improvisational elements were evident. For example, the wounded were delivered to hospitals using nearby buses rather than waiting for ambulances; and fourteen doctors and a nurse ran out of the headquarters of the British Medical Association to help the injured. In a dramatic demonstration of evolving technology, some enterprising passengers ('citizen journalists') e-mailed photos to the media from inside the subway tunnels, thereby scooping professional reporters who were prevented from reaching the disaster site by police.

Emergent elements in social movements and social movement organisations

While Turner and Killian's treatment of emergence was originally directed toward the situation of the acting crowd, later on they extended this to the case of social movements. In the third edition of their text, Turner and Killian (1987: 23) observed that in social movements 'the emergent normative element is the collective redefinition of a condition once viewed as a misfortune into a state of injustice'. It was further noted that 'this normative definition transforms what might otherwise be simple interest group politics into a crusade' (ibid). In other words, the development of a revised sense of justice is the keystone process in the development of a social movement and, as such, is central to the dual and interrelated processes of reconceiving reality and revising social norms (Turner 1981: 9). Finally, Turner and Killian make an important point that connects their work to more contemporary social movement theory. This sense of injustice as an emergent norm is not static. Rather, it 'motivates and crystallizes with the development of the movement' (1987: 243). This collective redefinition is a powerful tool both suffusing movement strategy and diffusing it outwards to sympathetic publics.

When the study of social movements rocketed in the 1980s and 1990s, Turner and Killian's emergence model was not on centre stage. Despite its divergence from the stereotypes of social movement genesis and participation found in classic collective behaviour theory, most in the newer generation of social movement scholars either were not acquainted with emergent norm theory or considered it outdated. This is unfortunate because the Turner and Killian perspective parallels in many ways the 'action–identity' theory of social movements that has been quite influential, especially in Italy and France.

One of the first researchers to note this was the American political scientist Sidney Tarrow (1988) who wrote that the 'constructionist' version of what had come to be known as 'New Social Movement Theory' runs in 'remarkable parallel to the earlier American emphasis on emergent norms; most notably with reference to the forging of new collective movement identities'. This is especially evident in the work of the late Italian psychotherapist and social movement theorist Alberto Melucci, who is credited with first coining the term 'New Social Movement' (NSM).

Turner (1981: 6) describes social movements as 'instrumental in the continuous construction and reconstruction of collective and individual views of reality', linking this to the symbolic interaction tradition in sociology. Melucci (1989: 25–6) too insists that a social movement is not a unitary empirical phenomenon, but rather a 'composite action system'. Individuals act collectively to *construct* their action by defining in cognitive terms new possibilities and limits, while concurrently interacting with others to organise (make sense of) their common behaviour. Along with ethnomethodology, phenomenology and expectation theory, Melucci cites symbolic interaction as centrally influencing his approach to social movements.[2]

Three 'emergent' elements characterise the growth of movements: new grievances, new collective identities, and new modes of association (Hannigan 1990).

Emergent grievances

In the Turner–Killian version of emergence theory, the development of a revised sense of justice is the cardinal process in the development of a social movement. Unfortunately, this has been somewhat of a blind spot in social movement research, partly reflecting the chronic absence of participant observers during the early stages of movement formation and growth. Some researchers have addressed this by attempting to recreate the dynamics of social movement formation in a controlled laboratory situation. In what is perhaps the best known of these studies, Gamson, Fireman and Rytina (1982) investigated how 'encounters with unjust authority' produced an emergent sense of opposition. They identified several 'classes' of protest activity: reframing (verbalisations of what is wrong), divesting acts (declarations of independence that sever people's obligation to authority), loyalty building, and internal conflict management.

More recently, there has been a spate of empirical studies focusing on the process of *consensus mobilisation*. The idea here is that social movement organisation (SMO) activities, goals and ideology will appeal more to potential recruits if they are congruent with existing interests, values and beliefs. Drawing on the work of Erving Goffman (1974), David Snow and his colleagues identified four processes that contribute significantly to consensus mobilisation: *frame bridging* (individual and social movement frames of reference already match: SMOs need only point out the similarities); *frame extension* (SMO frames are extended so as to align with the values and interests of potential adherents); *frame amplification* (an SMO frame is clarified and fortified through deliberately linking it to widely shared public values or beliefs); and *frame transformation* (individuals must have their world-views actively transformed in order to correspond with SMO frames).

As Rule (1989: 158) properly recognised, this framing approach is not all that different from the theoretical language for redefinitions of situations and normative innovations utilised by Turner and Killian. Dorcetta Taylor (2000: 511) describes this 'emergent' dimension of social movement framing in her essay on the environmental justice paradigm. Collective action frames, Taylor explains, are 'emergent, action-oriented sets of beliefs and meanings developed to inspire and legitimate social movement activities and campaigns designed to attract public support'. The word 'emergent here refers to the fact that these ideas, beliefs and norms are in the process of being formulated'. While collective action frames contain components of agency (empowerment) and identity construction, they are especially tied to the recognition of injustice and the articulation of grievances. Such frames are considered *injustice frames* because 'they are developed in opposition to already existing, established, and widely accepted frames'.

The major difficulty with this framing approach to social movement is that it perpetuates the assumption that normative sentiment is imposed or marketed by the SMO leadership rather than arising in a more organic fashion out of the self-reflexivity which is said to be characteristic of New Social Movements such as environmentalism.

Cable and Benson (1993) identify a new norm, that of *total justice* that emerged among community-based grassroots environmental organisations in the United States in the 1990s and played a crucial role in both their genesis and in the outcome of their activities. This norm 'encompasses two broad principles: *a general expectation of justice and a general expectation of recompense for loss and injuries*' [original italics]. While they observe that the emerging norm of total justice may be used as a frame for interpreting various pollution problems as grievances attributable to corporate wrongdoing, Cable and Benson are more inclined to see this developing organically among community activists rather than being 'marketed' to them by organisational leaders.

Emergent identities

According to Polletta and Jasper (2001: 285), *collective identity* is defined as 'an individual's cognitive, moral, and emotional connection with a broader community, category, practice or institution'. As such, it is conceptually and empirically different from personal identities, although it may form part of a personal identity. Poletta and Jasper identify four questions that prompted social movement scholars to theorise about collective identity. Why do collective actors come together when they do, particularly when their grievances are not readily visible? What persuades people to mobilise, especially in the absence of material incentives or coercion? Why do movements choose some strategic options over others? And, finally, how do changes in collective identity capture a dimension of social movement impact beyond those usually studied, most notably policy reform and expanded political representation? Together, these questions have suggested four distinct roles for collective identity in the emergence, trajectories and outcomes of social movements: the creation of collective claims, recruitment into movements, strategic and tactical decision-making, and movement outcomes (pp. 284–5).

One of the most concerted attempts to build a model of emergent identity formation was formulated by Melucci (1989: 35). He concluded that constructed collective identities involve three interwoven dimensions:

> First, formulating cognitive frameworks concerning the goals, means and environment of action; second, activating relationships among the actors, who communicate, negotiate and make decisions; and third, making emotional investments, which enable individuals to recognize themselves in each other.

Like Turner and Killian's revised description of the emergent norm, Melucci's understanding of collective identity formation is deliberately broad. Not only does it include the construction of a group identity, but it also entails an element of strategic action. Collective action, Melucci (1989: 25) states, is developed within a field of opportunities and constraints and thus is situated within an ever-changing environment. As such, it is compatible with the four distinct roles for collective identity suggested by Poletta and Jasper.

Most recently, the centrality of collective identities in understanding contemporary social movements has been challenged, specifically in relation to the 'anti-globalisation movement'. In these 'globalisation' conflicts, for example the 'actions' at Seattle, Genoa and Quebec City, it is 'not collective identity but new spaces of private experience that are increasingly at stake' (McDonald 2004: 590). This shift from 'role to experience' is depicted as an ongoing project wherein activists 'construct their own lives free of the constraints imposed by barriers of tradition, caste, order' (Dubet 2004: 707). While the participants broadly share a parallel critique of current economic and political structures and a shared vision for an alternative world (Faro 2004: 634–5), they only minimally embrace a common collective identity.

Emergent modes of association

Within the social movement literature, two basic forms of organisation have been identified: a centralised bureaucratic model and a decentralised, informal model. In purely strategic terms, each is optimally effective for a different task. However, as Jenkins (1983: 541) has noted, despite a broad historical shift towards more bureaucratised associations, decentralised movements have continued to emerge, especially in the case of redemptive or personal change movements. More recently, this decentralised mode of association has taken on an ideological character for many NSMs. As Keane and Mier (1989: 6) have noted, according to Melucci, the very focus of these movements – their patterns of interpersonal relationships and decision-making mechanisms – are valued as ends-in-themselves. As such, they are meant as deliberate signs or messages to the rest of society about how we should live in the future.

Given their prominent 'ecocentric' character, it should come as no surprise that many environmental SMOs have experimented with various decentralised forms of organisation.

Papadakis (1984) has traced the evolution of new forms of association among a variety of groups that form the understructure of *Die Grünen* (the Green Party) in Germany. When, in 1980, the Federal Government announced plans to build a reprocessing plant for nuclear waste in Gorleben, part of which was a nature reserve, ecological protestors not only staged a mass demonstration but 400 of them also remained on the site for an entire month and built an 'alternative village'. The inhabitants organised themselves into reference groups of up to 15 people, each of which sent a delegate to a 'Speaker's Council.' Papadakis describes various other forms of emergent modes of association. There were, for example, an estimated 11,500 'alternative projects' that provided services ranging from alternative technology enterprises to self-help therapy for squatters and 'alternative communities' where those disillusioned with the 'system' could temporarily find refuge from the technological society.

Some of these alternative groups embraced a 'consensus' mode of decision-making inspired variously by the Quakers and by aboriginal bands. For example, in the Clamshell Alliance, an anti-nuclear group that was active in the New England states in the late 1970s, no decision could be taken unless there was unanimous consent, although those who objected could 'stand aside'. While this had positive results for group solidarity, it ultimately split the Clamshell Alliance into two opposing factions: those who considered consensus decision-making as unduly constraining their capacity to act immediately and strategically; and the more egalitarian minded who regarded it as the cornerstone of their alternative approach (Barkan 1979; Downey 1986).

Social learning as an emergent process

Another central idea that percolated through the NSM literature in the 1980s was that of 'social learning'. This describes a process of collective reflection that informs and directs the collective action. Most authors linked it in some way with Jürgen Habermas' notion of 'communicative action' whereby social actors establish their interpersonal relations and coordinate their action by actively negotiating with one another and coming to an agreement. Social learning is said to involve the internal resolution of conflict through the successful practice of communicative action. The consensus decision-making of the Clamshell Alliance is said to be illustrative of communicative action in a social movement context.

One of the first explicitly to use the term social learning in an environmental context was the American political scientist Lester Milbrath. Milbrath described social learning (or 'social relearning', as he sometimes called it) as a shift from the Dominant Social Paradigm (DSP) to a New Environmental Paradigm (NEP), very much along the lines described by Catton and Dunlap. As long as our society is working reasonably well, he observed, the majority of the population is likely to tune out the environmental education efforts of scientists and social movements. Milbrath, who is both an optimist and tireless ecological proselytiser, was scarcely discouraged by this. We should not, he urged, perceive that a belief structure or paradigm is so firmly entrenched that it cannot be displaced (1989: 368). Ultimately, it will be nature itself that will be 'the most frequent spur to new thinking' and climate change is likely to be 'the most insistent and persistent teacher' (p. 376).

As we increasingly discover that technological fixes are not able to cope with escalating environmental problems, deep-set resistances to paradigm change will break down. In order to prepare for that day, Milbrath urged that we do everything we can to promote social learning, including re-orienting or re-designing our institutions so that they learn more readily and introducing a greater sense of spirituality into our lives (pp. 379–80). Rather than a collective endeavour, Milbrath viewed social learning more in terms of the convergence of millions of individual thoughts, desires and convictions among people who had begun to embrace an environmental consciousness.

Social learning has re-appeared as a motif in the literature on reflexive modernisation. In Beck's account, the risk defining process may best be understood as a plea for a form of social learning in which progress is redefined (Barry 1999: 162). This would occur in a zone located beyond the parameters of industrial society. Furthermore, the process of social learning crucially involves the collective acquisition of knowledge. Thus, Lipschutz (1996: 64) characterises social learning as a deliberate, incremental process of achieving consensual knowledge as it proceeds in the absence of absolute truth and is laden with arguments, uncertainties and contradictions.

VanWynsberghe (2001) adopts Lipschutz's definition of social learning in presenting a case study of a Heritage Centre operated by the Walpole Island First Nation, a Canadian aboriginal group. Here, he invokes the concept of a 'community-in-practice', a term from the organisational theory literature (see Wegner 1998) that explains how activist organisations build collective legitimacy and support by invoking shared historical frameworks and perspectives. The Heritage Centre, he found, was able to expand membership and retain support by engaging the native community in an ongoing process of social learning directed towards heritage issues.

Another contemporary environmental sociologist who has made extensive use of the concept of social learning is Robert Brulle (2000: 272–82). Brulle, a critical theorist in the mode of Habermas, claims that the ability of Americans to engage in social learning about the environment has been 'systematically blocked by the institutions of capitalism and the bureaucratic state'. The environmental movement is also culpable here for several reasons. First of all, environmentalists have failed to speak with a unified voice, engaging instead in a gaggle of 'multiple and partial discourses that are unable to appeal to a wide audience'. Second, mainstream environmental organisations have adopted an oligarchical structure that 'blocks citizen involvement, limits the range and scope of alternatives considered, and limits the organization's capacity for mobilization'. Like Milbrath, Brulle remains optimistic that these obstacles can be overcome. Social learning, he observes, depends on the creation both of alternative world-views and of social institutions that can translate and convey these into the public sphere. Environmental movement organisations (EMOs) are the logical candidates to serve as agents of social learning here. In order to do so, however, they must do two things. First, EMOs must adopt a more democratic structure by which members' participation is enhanced. Second, they must create an 'environmental "metanarrative" or "masterframe"' (Eder 1996: 207) which both unites the disparate discourses that currently fragment the US environmental movement and communicate with the American public in terms that make sense. Brulle does not spell out in any detail what this metanarrative would look like, although he suggests that it needs to combine scientific and legal competence with moral fervour and deep concerns over equity and justice. As an example of a partial metanarrative of nature that has enjoyed some success in recent years, he cites the term biodiversity, which he labels a 'discursive invention' (see Chapter 9 of this book).

Towards an emergence model of nature, society and environment

Drawing together strands from each of these varied literatures, it is possible then to begin to formulate an emergence model of nature, environment and society. One distinct advantage of this approach over other recent perspectives, notably various 'co-constructionist' models, is that it applies to a broader range of phenomena. Unlike actor–network theory and its various offshoots, it does not require inventing its own jargon. Finally, it is firmly rooted in the symbolic interaction tradition in sociology.

In sketching out the parameters of this emergence model, several key assumptions must be clarified.

First of all, our relationship with nature should be conceptualised as both fluid and emergent. This is not just a matter of fluctuating human perceptions and definitions, as social constructionists emphasise. Rather, we must also allow for the incorporation of materialist elements, what Lockie (2004: 26) terms the 'substance and patterns of nature'. Note, however, this should not become a kind of Trojan horse for the ecocentrism that runs deeply through so much contemporary analysis of nature and society. In other words, to recognise that the relationship between the social and the material is both interactive and emergent does not automatically validate the claim that the agency of nature is fully equivalent to the agency of humans.

One important conduit for a more dynamic view of nature and society is through the process of *improvisation*. Thus, in discussing the 1978 ice storm that severely impacted large sections of Eastern Canada and the Northeastern United States, Murphy (2004: 11) observes that such severe disturbances of nature act as a *prompt*, influencing human conceptions, discourses and practices and inciting an 'improvised response'. And, improvisation, as we have seen, is closely identified with Shibutani's (1966) description of rumour transmission as 'improvised news'.

Key dimensions

Emergent uncertainties

Whether or not life today is any more hazardous than it was a century ago, it often appears that way. Mad cow disease (BSE), AIDS, SARS, the avian flu, 'Frankenfoods', global warming, nuclear accidents – each week seems to herald the arrival of some new danger. What all of these have in common is an overpowering sense of ambiguity or contingency, what Sartre (1975: 100) once called 'the vertigo of possibility' (cited in Horlick-Jones

2004: 108). Even as the incidence of known infectious diseases declines (at least in Western nations), new uncertainties arise (Zinn 2005: 1) both in relation to degenerate and chronic diseases such as multiple sclerosis and to unidentified viral infections. Frequent episodes of 'strange' or 'unprecedented' weather signify that something unusual is happening to the Earth's environment, but it is difficult to know exactly what. In short, the relationship between nature and humans is becoming more complex and more indeterminate (Wynne 2002: 471–2).

Congruent with Beck's risk society thesis, many sociological analysts argue that it is techno-science that must be held accountable for this upsurge of uncertainty since it has churned out an escalating flow of new synthetic chemicals, bioengineered foods and animal clones without ever stopping to consider their interactive effects or long-term impacts. One particular source of acute uncertainty are novel and innovative 'hybrids' such as stem cells and xenografts that involve the transplantation of tissues between different species and that hover controversially on the horizons of medical innovation (Brown and Michael 2004).

Beck and his acolytes have linked this climate of uncertainty to the 'demystification' of science and the eroding power of experts. As the authority of science wanes, they argue, ordinary people are exposed to a bewildering array of conflicting claims and discourses. As Murphy and Maynard (2000: 134) observe, the 'more uncertain the facts of the issue, the more prone it is to be socially constructed'. They cite as an example the debate over genetic testing – novel, lacking in precedents, and fraught with uncertainties. In such disputed arenas, the policy-making process is particularly vulnerable to 'rhetorical influences'.

Barbara Adam (1996: 95–7)) observes that time is centrally implicated in the emergence of this 'prevailing uncertainty' for four reasons. First, past knowledge has consistently proven to be of limited value in predicting a future that is characteristically indeterminate and contingent. To a considerable extent, this can be traced to the nature of scientific innovation. A quarter-century ago, for example, few could have imagined the changes induced by the widespread adoption of cell phones and other personal communication devices. Second, cycles of innovation and obsolescence have shrunk drastically, as have globalised hazards. This creates 'out-of-sync time-frames' in which the negative effects of new technologies such as nuclear power plants appear almost immediately, making them instantly obsolescent. Third, modern technologies are designed as isolated, bounded units that stand apart from the environment. In reality, they produce effects that become integrated into the ecological web. Furthermore, 'machine time' is reversible (as for example in video-recorders or DVDs) , while ecological time is not. Thus, there is no reversing the harmful effects of dumping toxic wastes into the groundwater. Finally, the links between cause and effects become obscured in late modernity. With some environmental hazards, there is a time-lag during which no visible symptoms emerge. We are no longer dealing with static, isolated phenomena but with interconnected, continuously changing, dynamic situations in which the link between input and output is far more complex than pollutants pouring out of a pipe into a stream. In the case of global warming, for example, cause and effect do not emerge in a linear manner, making this an environmental issue 'replete with uncertainties and the prospect of an indeterminate future' (p. 97).

This sense of uncertainty and indeterminancy is crucial to emergent theory because it leaves people without a firm set of cognitive guidelines. Like disaster victims or those caught up in civil disturbances, citizens today are stranded in a twilight zone.

Emergent organisation and structure

Ever since the potential threat of significant damage to nature and the environment first impacted public consciousness in the late nineteenth century, we have witnessed the appearance of new varieties of organised response. As is the case in disaster situations, some of these have been extensions or expansions of existing structures, others have displayed a more emergent character. Most have resided within the institutional boundaries of science, industry and government, but some have germinated and grown outside in what has become known as 'civil society'.

Frank (1997; *et al.* 2000; 2002) has written extensively on the emergence, challenge and eventual triumph of a 'scientific' model of nature protection over the formerly prevalent 'humanitarian' one. He explains this by changes in the predominant cultural and organisational frame. Specifically, nature protection is said to have passed through three main stages of global institutionalisation: change in world culture; change in world organisation; and change in nation-state politics. While nature protection was eventually incorporated into the society action system 'in highly specific, recipe-like ways' (Frank 2002: 49), this was initially a more emergent process in which the outcome was by no means given.

More recently, a body of research has been published that focuses on *emergent boundary organizations*, so named because they lie on the boundary between politics and science (Guston 2000). For example, Agrawala *et al.* (2001) have described the history of the International Research Institute for Climate Prediction (IRI), a boundary organisation created in 1996 to help coordinate, conduct, implement and evaluate research on seasonal climate variations and their impacts. The IRI is 'situated between the relatively different social worlds of climate modeling and forecasting on the one end, and agricultural, health, and other social and political decision making on the other' (p. 471). As such it operates not only in the scientifically uncertain milieu of predicting a major meteorological and oceanographic event such as an El Niño but also on a global political stage where the global climate change debate continues to rage.

Miller (2001) has developed the concept of *hybrid management* better to understand how boundary organisations such as the IRI function in international politics. More specifically, he examines in detail the processes by which they are constructed, taken apart and ordered in relation to one another. Taking as an example, the SBSTA (Subsidiary Body for Scientific and Technological Advice), a forum created in 1992 by the UN Framework Convention on Climate Change (UNFCCC), Miller identifies four elements that make up the process of hybrid management: hybridisation, deconstruction, boundary work and cross-domain orchestration. Contrary to Latour's (1993) thesis that the basic drive of modernity has been to purify hybrids, Miller suggests that boundary organisations such as this exist in order to 'establish and maintain a productive tension between the multiple, diverse forms of life in contemporary societies' (p. 487). Furthermore, power relationships within the SBSTA were found to be complex, with neither scientists, nor government, nor a single country such as the US able to monopolise the production of methods for measuring the emissions of greenhouse gases.

Emergent flows

Finally, the society–environment relationship may be conceptualised in terms of 'emergent flows'.

Drawing on the works of Manuel Castells and John Urry, Mol and Spaargaren (2003) have proposed a new 'sociology of environmental flows'. The central inspiration for this is the rapid social and economic globalisation of the planet and the increasing complexity that accompanies it. This is said to have rendered national states extinct and local 'places' and their inhabitants vestigial. Typically, data files begin the day in India, are forwarded to programmers in New York and Chicago and are finished off in California. Corporate executives are in perpetual motion, as likely to spend the majority of their time in airport lounges and international hotels as in the city where they nominally reside. Refugees and other immigrants travel on a global conveyor belt. Cities become little more than 'nodes' in global economic networks. Power relations are transformed. As Castells (1996: 412) memorably phrases it, 'the power of flows takes precedence over the flows of power'.

The nascent 'sociology of flows' upon which Mol and Spaargaren model their paper takes this even further. In Urry's (2003) view, both social actors and nation states fade away, ceding the stage of modernity to globally integrated networks and flows. Urry merges this idea of flows with the notion of 'hybridity', which, as we have seen, is central to actor–network theory. Thus, material objects and social relations dovetail to the point where they are indistinguishable. Finally, Urry adds a sprinkle of 'complexity theory' wherein outcomes develop in unpredictable, non-linear directions, and even chaotic directions. This puts the spotlight on the 'emergent properties' of global flows and networks.

Mol and Spaargaren point out that environmental sociology and the environmental sciences have long been concerned with flows. Mostly this has been centred on flows of material substances and energy or, more recently, in the flows of pesticides and other pollutants through ecosystems. However, Allan Schnaiberg's model of the 'treadmill of production' with its focus on human additions to the natural environment (causing pollution) and withdrawals (causing depletion) can also be seen as relevant (p. 12). This, however, is said to be too narrow, static and localised. An environmental flow, they claim, 'is not only or just material substances and technical infrastructures, but also the scapes, nodes, networks and discourses which go along with the flows or fluids in question' (p.17).

Most of what has been written about this sociology of environmental flows is pitched to an abstract level. One notable exception is an article by Harris Ali and Roger Keil (forthcoming) on global cities and the spread of infectious disease.

In their case study, Ali and Keil examine the transmission and response to an outbreak of SARS (Severe Acute Respiratory Syndrome) in 2003, arguing that this is fully understandable only in the context of the type of global networks (Castells) and flows (Urry) that we have been discussing.

On 5 March 2003, a 78-year-old Chinese-Canadian woman who had just returned home from a visit to relatives overseas died in a Toronto hospital, followed two days later by her son. She had evidently contracted the SARS virus during a stay in the Metropole hotel in Hong Kong. At least twelve guests at the Metropole were evidently 'infected' by a professor of medicine who had come to Hong Kong for a family wedding, after having treated SARS-infected patients in Guangzhou. Within the month, there were 13 SARS related deaths, 97 probable cases and 1,137 suspected cases in the Toronto area. These figures were even higher in Hong Kong and China. A second outbreak in Toronto in May 2003 led to five patients being quarantined.

Ali and Keil (2005) demonstrate that the pace and patterns of viral transmission reflect the 'time–space compression' (Harvey 1989) resulting from today's globalising forces. One important factor is the combination of a reduced incubation period (for the SARS virus this is only 2–10 days) and increased aeroplane travel between the nations and cities of the world. This has meant that travellers will likely have already returned home before the telltale symptoms appear. Second, they cite the proliferation of 'diaspora' or 'transnational' communities, especially in global cities. These are made possible by modern technologies of transportation and communication. Whereas in the great migrations to America from Central and Southern Europe in the early twentieth century, the immigrant might never return unless he or she acquired wealth, today airports and cell phone networks are clogged with people keeping in touch with relatives in the old country. This impacted Toronto on several occasions during the SARS emergency. A nurse's aid who had become infected through contact with her roommate's mother (and later died) travelled to Manila and attended a large wedding. Also, members of a charismatic Catholic religious group, most of Filipino descent, attended the funeral of an elderly member who had contracted SARS, early on, leading to the imposition of a mass quarantine.

Emergence plays a role here in several ways.

First, SARS is one, but not the only example of an emerging and spreading infectious disease. It is worth noting that three of the books cited in Ali and Keil's bibliography (*Secret Agents: The Menace of Emerging Infections* (Drexler 2003); *The Coming Plague: Newly Emerging Diseases in a World Out of Balance* (Garrett 1994); *Emerging Viruses* (Morse 1993)) use this terminology. These 'newly emergent and resurgent diseases' reflect changes in the human–environment relationship that encourage cross-species transfer. For example, in the SARS outbreak this probably occurred in the unsanitary, live animal markets of Guandong Province where 'exotic' animals (in this case, the civet cat) are sold as food delicacies.

Second, local–global interactions produce a series of 'unexpected, disproportionate and emergent effects' (Ali and Keil forthcoming). This ties in to the 'complexity' dimension that is associated with emergence. In this case, it is not only human relationships that are non-linear but also the interaction of pathogens with economic, political and social factors in unanticipated ways.

Conclusion

To propel the analysis of societal–environmental relations forward into new territory, it is necessary, to borrow a phrase from Steve Yearley (2002b), to undertake a series of 'Herculean labours'.

First of all, any fresh attempt to conceptualise environment-related matters needs to confront the 'nature–society divide'. This has unfortunately become somewhat of a fetish in contemporary environmental scholarship, occluding the pursuit of other theoretical ventures. Nonetheless, I think it is as unwise to bracket out the natural from sociological analysis, as it is to deny that nature and the environment are socially constructed. Coming to grips with this nature–society dualism is an especially daunting labour, further complicated by the increasing proliferation of 'hybrids' and 'cyborgs' whose constitution is, materially at least, part social, part natural.

By engaging in this first task, we inevitably encounter a second. Recent efforts to bridge the nature–society divide have thus far proven somewhat empirically elusive, operating as they do on a rather abstract plane. As discussed in Chapter 2, purveyors of actor–network theory and other co-constructionist approaches deliberately adopt an idiosyncratic jargon and conceptual repertoire that does not easily engage with any of the major theoretical perspectives in contemporary sociology. This is true even with more empirically grounded pieces of research (Swyngedouw 1999; 2005) that document the historical production of the 'socionatural'. Urry's 'sociology of flows', which partly incorporates ANT thinking, does much the same, with its jargon of attractors, iteration, chaos and equilibrium, autopoiesis, fluxes and time as a nominator of *dx/dt*. Mol and Spaargaren justify this on the grounds that a complexity-based 'socionomy as the new disciplinary hybrid' may be the only way to make sense of the dynamics of globalisation that otherwise are 'beyond systematic analyses and understanding' (p. 9).

Dunlap (2002b: 16) predicts that 'the future will see the emergence of new efforts to analyse societal–environmental relations that reflect a synthesis of the strengths of both [the conceptual and the empirical] streams, and this can only benefit our field'. As Martha Stewart is prone to say, this is 'a good thing', but right now it remains an only slightly less Herculean labour than closing the nature–society divide.

Third, and not unrelated to the second labour, is the challenge of reconciling macro-level, European-style, sociological theorising on the environment with the more particularistic data analysis characteristic of American environmental sociology. As Buttel (2002: 52–3, footnote 8) points out, 'a large share of the sociology papers on the environment published in Europe essentially consists of pieces of cultural sociology or research on social movements that happen to consider ecology and related movements to be indicative of interesting types of modern social movements'. On the other hand, American environmental sociology is more directly concerned with inequalities related to race, class and gender. As Fisher (2003:10) notes, this line of thought (especially in its theoretical version) 'tends to consider environmental problems to be a relatively direct consequence, or at least a clear correlate of industrialization and capitalist accumulation'.

In this chapter, I have suggested that an emergence framework can be a useful tool in undertaking these Herculean labours. One major advantage here is that it allows a range of phenomena – infectious diseases, ice storms and tsunamis, uncertainties and risks, scientific boundary organisations, environmental movements – to be conceptualised within the same framework. While it incorporates flow processes associated with globalisation (as in the Ali and Keil study of SARS), emergence theory is equally useful at the local level, for example in accounting for social interaction in the aftermath of disasters. Although it implies a 'bottom up' model of social learning, emergence theory is not explicitly prescriptive. Finally, as per Dunlap's prediction, it allows a synthesis of the theoretical and the empirical in a more seamless manner than do other contemporary approaches, most notably actor–network theory.

Notes

1 Environmental sociology as a field of inquiry

1. Addressing this point directly, Sorokin comments, 'A reader of these lines may think Dr Huntington has at his disposal there the detailed record of the Meteorological Bureau of Ancient Rome' (1964 [1928]: 191).
2. This view is not, however, universally shared. For example, Goldblatt (1996: 3) states 'of the classical trinity [Durkheim, Weber, Marx], Weber's work conducts the most limited engagement with the natural world'.
3. As it happens, NEP originally stood for 'New Environmental Paradigm', but Catton and Dunlap renamed it in 1980 in recognition of the increasingly ecological perspective involved in most environmental research (Freudenburg and Gramling 1989: 445). The HEP/NEP model was first briefly introduced in Dunlap and Catton (1983) and was elaborated in several papers presented at scholarly meetings in the late 1980s (see Dunlap 1993: 734–5).

2 Contemporary theoretical approaches to environmental sociology

1. I am grateful to Filip Alexandrescu for this insight and the references from Boulding's work in the 1950s.
2. This encompasses Schnaiberg and his former doctoral students, Kenneth Gould, David Pellow and Adam Weinberg.
3. Spaargaren (2000: 64–5) takes umbrage at this statement, declaring the optimism–pessimism dichotomy to be less than helpful. Science and technology, he argues, are important vehicles in the ecological modernisation process, but this 'does not imply, however that one would automatically or inevitably lapse into a technological fix approach'. This is more the case for a 'strong ecological modernisation' which purports to be more open to broad-ranging changes to society's institutional structure and economic system than for 'weak ecological modernisation' that emphasises technological solutions to environmental problems and 'looks like a discourse for engineers and accountants' (Dryzek 2005: 172–3).

3 Environmental discourse

1. This contrasts with Dryzek (2005: 10) who clearly spells out his intent to 'lay out the basic structure of discourses that have dominated recent environmental politics' and to 'produce something more than just an account of environmentalism'.
2. Grant is one of the more controversial figures in the early wilderness protection movement. A patrician lawyer with close links to many élite figures in business and politics including Teddy Roosevelt, he was among other things a founder of the Save the Redwoods League, the New York Zoological Society and the Boone and Crocket Club. At the same time, he has been called by historian John Higham (1963) 'intellectually the most important nativist in recent American history'. Grant's book, *The Passing of the Great Race* (1921), was for a while a popular-selling exposition on the principles of eugenics although it was less successful in subsequent printings. Grant's concern with the

subject of eugenics and racial exclusion was shared by a number of other leading wilderness protectionists of the day including William Hornaday, Fairfield Osborn and Vernon Kellogg.

3 On December 19, 1913 US President Woodrow Wilson signed legislation that permitted the construction of a dam across the Hetch Hetchy Valley in California's Yosemite National Park. The dam and accompanying reservoir allowed water from the Tuolumne River to be diverted to San Francisco, thereby supplying that city with both drinking water and hydroelectric power. The project was strenuously but unsuccessfully opposed by John Muir and the Sierra Club on the grounds that it would destroy the natural beauty of the Valley and make it an unsuitable habitat for wildlife. See Magill 1995: 106–9.

4 Anna Sewell's book *Black Beauty* was originally published in England in 1877 where it sold more than 90,000 copies. It was republished by the American Humane Education Society in 1890. While designed to increase support for the animal welfare movement, the book also helped to establish a climate for the wider support of wildlife conservation (Lutts 1990: 22–3).

5 In the 1930s, due largely to the efforts of Charles Adams, director of the New York State Museum and Paul Sears, a plant ecologist, The Ecological Society in the US did make some attempt to bring social scientists and ecologists together in a common forum, notably in a joint symposium of the Society with the American Association for the Advancement of Science entitled 'On the relation of ecology to human welfare – the human situation'. Sadly, two of the leading theorists of the Chicago School, Ernest Burgess and Roderick McKenzie, were unable to attend, leaving August Hollingshead as the only representative of sociology (Cittadino 1993).

6 Kwa (1993: 248) dates the beginning of 'ecosystem ecology' to 1953 when Eugene Odum, a University of Georgia zoologist, published his influential text, *Fundamentals of Ecology*. Soon after, Odum began a series of radioecological studies under the sponsorship of the Atomic Energy Commission (AEC) to determine the impact on the environment of a new atomic weapons plant that was to be built on the Savannah River in South Carolina.

7 The term 'organisational weapon' was first introduced in Philip Selznick's classic (1960) study of the American Communist Party. Eyerman and Jamison (1989) borrow the concept to describe Greenpeace's use of flamboyant and sometimes illegal media-capturing actions to pressure governments and business. Organisations are weapons in such cases when they act in a manner that is considered unacceptable by the community.

8 In a 1992 interview, Lois Gibbs, the heroine of the Love Canal story, told environmental activist and author Robert Gottlieb: 'Calling our movement an environmental movement would inhibit our organizing and undercut our claim that we are about protecting people, not birds and bees' (Gottlieb 1993: 318).

9 In an article published posthumously, Chavez (1993: 166–7) charges that corporate growers in California effectively sidestepped many of the provisions of these contracts, including those governing the use of pesticides. Chavez observes that many of these same growers were the largest financial contributors in the campaign to defeat Proposition 128 (nicknamed 'Big Green'), a 1990 ballot initiative supported by environmental groups and the UFW which among other things would have 'protected California's last strands of privately held redwoods and banned cancer-causing pesticides'.

10 The term 'environmental racism' was evidently coined by the Reverend Benjamin Chavis, former head of the United Church of Christ Commission on Racial Justice and later Executive Director of the National Association for the Advancement of Colored People (NAACP), a major civil rights organisation in the United States (Higgins 1993: 287).

11 The impetus for this study was a request from Walter Fauntroy, a congressional representative from Washington, DC, and an active participant in a struggle in Warren County, North Carolina, to stop the establishment of a toxic landfill containing PCB-laced soil (Bryant and Mohai 1992: 2).

12 It should be noted that the funding for this conference was gold-plated, including among other sources, the Ford Foundation and the Rockefeller Family & Associates (Mayer 1992).

5 Social construction of environmental issues and problems

1 Ibarra and Kitsuse (1993) also outline a set of 'counterrhetorical strategies' which are meant to block claimants' attempts to construct a problem and/or demand action.

2 This was suggested at the public hearings on the proposed Alberta–Pacific bleached Kraft pulp mill in Northern Alberta by Cindy Giday from the Northwest Territories who was the lone native (and female) on the Alpac EIA Review Board.
3 Total membership of the twelve or so major national environmental organisations in the US increased from about four million in 1981 to roughly seven million in 1988 (Bramble and Parker 1992: 317).
4 Note, however, that in the course of fundraising and lobbying, major conservation organisations are inclined to draw from both rhetorics, anchoring their appeals in both commercial and moral rationales (Yearley 1992: 26).
5 Unfortunately, the *sao la* soon faced extinction as collectors from around the world attempted to obtain one, even reputedly offering a bounty of up to $1 million (Shenon 1994).

6 Media and environmental communication

1 The phrase 'Spaceship Earth' was evidently coined by the British economist Barbara Ward as the title of a book she published in 1966 on the links between economics and the environment (Pearce 1991: 11).
2 The only exception to this was the *New York Times* coverage that continued to separate various aspects of environmental issues.
3 Reporters' first choice here is usually a government spokesperson rather than a scientific expert. Sandman *et al.* (1987) suggest that one major reason for this is that reporters generally want two very specific types of environmental risk information: how much of the hazardous substance is in the air or water and how much of this substance does it take to cause problems.
4 Nearly twenty years earlier, an American researcher (Witt) noted a similar diversity of environmental sources. Witt's results indicated that the primary news sources of environmental reporters were conservation clubs and organisations followed closely by business and industry sources. It is worth noting that unlike Cottle, Witt did not extract his sources from media content alone, relying instead on a national questionnaire survey of environmental reporters working for US newspapers.
5 Einsiedal and Coughlan (1993) found some revealing differences when they compared the environmental content in Canadian daily newspapers with full-time environmental writers with that in papers that utilised general reporters. On the whole, there were more environmental stories in the former; the environmental beat reporters were more likely to write longer, more analytical, self-initiated pieces and they were more likely to challenge conventional institutional wisdom.
6 According to a survey carried out by *Editor & Publisher* in the summer of 1970, there were 107 environmental reporters working in the American media, mainly on daily newspapers (Schoenfeld 1980: 456).
7 I witnessed this firsthand when doing observation in the newsroom of a national television network in Canada. One day, a senior producer was visibly upset when he received a letter from a viewer charging that the national news broadcast had been giving too much time to an anti-nuclear protest despite the newsworthiness of the issue (see Hannigan 1985).
8 Not coincidentally, perhaps, the Lewingtons' daughter, Jennifer, is a veteran beat reporter with the (Toronto) *Globe & Mail*.
9 'Monkey wrenching' or 'ecotage' refers to a wide range of actions by radical environmental activists to disrupt and halt damage to the environment including pouring abrasives into the crankcases of road-building vehicles, pulling up surveyors' stakes and 'spiking' trees by driving long metal spokes into them. The name comes from Edward Abbey's 1976 novel, *The Monkey Wrench Gang*, in which a group of ecoteurs plot to blow up the Glen Canyon Dam (see Franck and Brownstone 1992: 190; Manes 1990: 8–9).

7 Science, scientists and environmental problems

1 An exception to this is Germany where the precautionary principle has been enshrined historically.
2 Scientific concern over pesticide poisoning began more than two decades prior to the publication of *Silent Spring*. As far back as 1945, Rachel Carson herself evidently attempted unsuccessfully to interest *Reader's Digest* in commissioning an article from her on the research being conducted by colleagues at

the Paxutent Research Center indicating that the pesticide DDT had adverse effects on the reproduction and survival of birds after repeated applications (Lear 1993: 33). In the early 1950s, an emerging consensus in the US public health field that the use of chemicals in food production needed to be more strictly regulated led to 46 days of Congressional hearings. However, the issue was seen as narrow and technical and received little media attention. Unlike the eventual environmental campaign sparked by Carson's book, evidence that pesticides might cause harm somewhere down the road was not as compelling to the media and the mass public as dramatic images of dead birds (Bosso 1987: 80).

8 Risk

1 For many years, human sewage from many local households mixed together with storm-water in the same pipe. There has since been a vigorous sewage separation programme, but some homes and businesses still discharge sewage into the storm-water system.
2 One exception to this is a 1984 decision in the *Ferebee v. Chevron Chemical Co.* case in the United States that allowed the jury to rely on the testimony of individual physicians in the absence of ironclad epidemiological evidence concerning injury by exposure to pesticides (see Cronor 1993).
3 As it happens, the Review Board recommended that the mill should not be built unless further studies indicated that it would not pose a serious hazard to biological life in the river and for downstream users along the Peace–Athabasca river system. Nine months after it agreed to abide by these findings, the Alberta government overturned its own decision and decided to allow Alpac to proceed.

9 Biodiversity loss: the successful 'career' of a global environmental problem

1 At the first official SCB meeting in April, 1988, many participants cited the need for aggressive conservation action rather than research as the top priority (Tangley 1988: 444).
2 By contrast, taxonomy, a speciality science that involves identifying and cataloguing biological species, has been in steady decline for decades. Perceived to be a nineteenth-century descriptive science with little present-day application (Burton 2003), taxonomy was unable to claim ownership of biodiversity as an environmental problem, despite its vital importance in compiling species lists.
3 My chronology of these international conventions draws primarily on 'Annex 3: international legislation supporting conservation of biological diversity' in McNeely *et al.* (1990a).
4 The Northern spotted owl became one of the 'most celebrated and vilified endangered species' (Grumbine 1992: 144) in recent memory. With a habitat and geographic range that stretches the length of old growth forests from British Columbia to Northern California, protecting it under the Endangered Species Act implied a significant reduction in logging activities in the ancient forests. In the course of a decade of political and legal wrangling the Northern spotted owl became a symbol for some of the unrealistic features of the Act.
5 This appears to have been a two-way street. Not only did the fate of the dinosaurs provide a powerful magnet by which diversity activists could attract the attention of the public, but also research on the immediate threat of extinction has proven useful in understanding what happened 245 million years ago. For example, Niles Eldredge, in writing his book *The Miner's Canary: Unraveling the Mysteries of Extinction* (1991) relied heavily on Edward Wilson's published data and arguments to examine the relationship between the mass extinctions of the geological past and the present-day biodiversity crisis (Eldredge 1992: 90).

10 Towards an 'emergence' model of environment and society

1 In the rebel province of Aceh, by contrast, a peace settlement was tentatively announced in July 2005 between the Indonesian administration and the Acehnese government-in-exile in Sweden.
2 In the summer of 1990, I presented a paper outlining these parallels (see Hannigan 1990) at the World Congress of Sociology in Madrid. Fortuitously, both Alberto Melucci and Ralph Turner were in attendance. Privately, both confirmed that there were similarities. Interestingly enough, Turner told me that he and Melucci had not previously met until the week before, when they both participated in a conference in Berlin.

Bibliography

Adam, B. (1996) 'Re-vision: the centrality of time for an ecological social science perspective', in S. Lash, B. Szerszynski (eds) *Risk, Environment and Modernity: Towards a New Ecology*, London: Sage.

Adams, W.M., Aveling, R., Brockington, D., Dickson, B., Elliott, J., Hutton, J., Roe, D., Vira, B. and Wolmer, W. (2004) 'Biodiversity conservation and the eradication of poverty', *Science*, 306 (No. 5699): 1146–9.

Aglionby, J. (2005) '"Ghosts" seek their pre-tsunami past', *Guardian Weekly,* 24–30 June: 10.

Agrawala, S., Broad, K. and Guston, D. H. (2001) 'Integrating climate forecasts and societal decision-making: challenges to an emergent boundary organization', *Science, Technology & Human Values,* 26(4): 454–77.

Alexander, J. C. and Smith, P. (1986) 'Social science and salvation: risk society as mythical discourse', *Zeitschrift für Soziologie*, 25: 251–62.

Ali, H. and Keil, R. (forthcoming) 'Global cities and the spread of infectious disease: the case of severe acute respiratory syndrome (SARS) in Toronto, Canada', *Urban Studies*.

Alpert, P. (1993) 'Support for biodiversity research from the U.S. Agency for International Development', *BioScience*, 43(9): 628–31.

Altheide, D. (1976) *Creating Reality: How TV News Distorts Events*, Beverly Hills, CA: Sage.

Anand, R. (2004) *International Environmental Justice: A North–South Dimension*, Aldershot: Ashgate.

Anderson, A. (1993a) 'Source–media relations: the production of the environmental agenda', in A. Hansen (ed.) *The Mass Media and Environmental Issues*, Leicester: Leicester University Press.

Anderson, A. (1993b) *Media, Culture and the Environment*, New Bruswick, NJ: Rutgers University Press.

Angier, N. (1994) 'Redefining diversity: biologists urge look beyond rain forests', *New York Times,* 29 November: B–5; B–9.

Aronson, N. (1984) 'Science as a claims-making activity: implications for social problems research', in J. Schneider and J. I. Kitsuse (eds) *Studies in the Sociology of Social Problems*, Norwood, NJ: Ablex.

Asch, S. E. (1951) 'Effects of group pressure upon the modification and distortion of judgment', in H. Guetzkow (ed.) *Groups, Leadership and Men*, Pittsburgh, PA: Carnegie Press.

Bakker, K. (2002) 'From state to market? Water *mercantización* in Spain', *Environment and Planning A*, 34(5): 767–90.

Balch, A. and Press, D. (2003) 'The politics of biodiversity: a political case of the endangered species act', in S. L. Spray and K. L. McGlothlin (eds) *Loss of Biodiversity*, Lanham, MD: Rowman & Littlefield.

Barkan, S. (1979) 'Strategic, tactical and organizational dilemmas of the protest movement against nuclear power', *Social Problems*, 27: 19–37.

Barry, J. (1999) *Environment and Social Theory*, London: Routledge.

Barton, J. H. ((1992) 'Biodiversity at Rio', *BioScience*, 42(10): 773–6.

Baumann, E. A. (1989) 'Research rhetoric and the social construction of elder abuse', in J. Best (ed.) *Images of Issues: Typifying Contemporary Social Problems*, New York: Aldine de Gruyter.

Beck, U. (1992) *The Risk Society: Towards a New Modernity*, London: Sage.

Benedick, R. E. (1991) *Ozone Diplomacy: New Dimensions in Safeguarding the Planet*, Cambridge, MA: Harvard University Press.

Benford, R. D. (1993) 'Frame disputes within the nuclear disarmament movement', *Social Forces*, 71(3): 677–701.

Benton, T. (1991) 'Biology and social science: why the return of the repressed should be given a (cautious) welcome', *Sociology*, 25(1): 1–29.

Best, J. (1987) 'Rhetoric in claims-making', *Social Problems*, 34(2): 101–21.

Best, J. (1989) 'Afterword: extending the constructionist perspective: a conclusion – and an introduction', in J. Best (ed.) *Images of Issues: Typifying Contemporary Social Problems*, New York: Aldine de Gruyter.

Bierstedt, R. (1981) *American Sociological Theory: A Critical History*, New York: Academic Press.

Blakeslee, A. M. (1994) 'The rhetorical construction of novelty: presenting claims in a letters forum', *Science, Technology & Human Values*, 19(1): 88–100.

Bloor, D. (1999) 'Anti-Latour', *Studies in the History and Philosophy of Science*, 30: 81–112.

Blowers, A. (1993) 'Environmental policy: the quest for sustainable development', *Urban Studies*, 30(4/5): 775–96.

Blowers, A. (1997) 'Environmental policy: ecological modernization or the risk society?' *Urban Studies*, 34(5/6): 845–71.

Blowers, A., Lowry, D. and Solomon, B. D. (1991) *The International Politics of Nuclear Waste*, London: Macmillan.

Blühdorn, I. (2000) 'Ecological modernization and post-ecologist politics', in G. Spaargaren, A. P. J. Mol and F. Buttel (eds) *Environment and Global Modernity*, London: Sage.

Bocking, S. (1993) 'Conserving nature and building a science: British ecologists and the origins of the Nature Conservancy', in M. Shortland (ed.) *Science and Nature: Essays in the History of the Environmental Sciences*, Oxford: British Society for the History of Science.

Boehmer-Christiansen, S. (2003) 'Science, equity and the war against carbon', *Science, Technology & Human Values*, 28(1): 69–92.

Bora, A. and Dobert, R. (1992) 'Prerequisites of procedural rationality: notes on a German technology assessment of transgenic plants with herbicide resistance', paper presented at the Symposium 'Current Developments in Environmental Sociology', International Sociological Association, Woudschoten, The Netherlands.

Bosso, C. J. (1987) *Pesticides and Politics: The Life Cycle of a Public Issue*, Pittsburgh, PA: University of Pittsburgh Press.

Boulding, K. E. (1950) *A Reconstruction of Economics*, New York: Wiley.

Boyne, R. (2003) *Risk*, Buckingham: Open University Press.

Bramble, B. J. and Parker, G. (1992) 'Non-governmental organizations and the making of U.S. international environmental policy', in A. Hurrell and B. Kingsbury (eds) *The International Politics of the Environment*, Oxford: Clarendon Press.

Bramwell, A. (1989) *Ecology in the 20th Century: A History*, New Haven, CT: Yale University Press.

Braun, B. and Castree, N. (1998) *Remaking Reality: Nature at the Millennium*, London: Routledge.

Bray, D. and von Storch, H. (2005) 'Survey of climate scientists 1996, 2003', Online: http://w3g.gkss.de/G/mitarbeiter/bray.html/BrayGKSSsite/BrayGKSS/surveyframe.html

Breyman, S. (1993) 'Knowledge as power: ecology movements and global environmental problems', in R. D. Lipschutz and K. Conca (eds) *The State and Social Power in Global Environmental Politics*, New York: Columbia University Press.

Broad, W.J. (2005) 'With a push from the U.N., water reveals its secrets', *The New York Times*, 26 July: D1.

Brookes, S. H., Jordan, A. G., Kimber, R. H. and Richardson, J. J. (1976) 'The growth of the environment as a political issue in Britain', *British Journal of Political Science*, 6: 245–55.

Brouilette, J. R. and Quarantelli, E. L. (1971) 'Types of patterned variation in bureaucratic adaptations to organizational stress', *Sociological Inquiry*, 41: 39–46.

Brown, L. R. (1986) 'And today we're going to talk about biodiversity...that's right, biodiversity', in E. O. Wilson (ed.) *Biodiversity*, Washington, DC: National Academy Press.

Brown, N. and Michael, M. (2004) 'Risky creatures: institutional species boundary change in biotechnology regulation', *Health, Risk & Society*, 6(3): 207–22.

Brulle, R. J. (1998) 'Review of Hannigan (1995)', *Sociological Inquiry*, 68(1): 137–9.

Brulle, R. J. (2000) *Agency, Democracy and Nature: The U.S. Environmental Movement from a Critical Theory Perspective*, Cambridge, MA: MIT Press.

Bryant, B. and Mohai, P. (1992) 'Introduction', in B. Bryant and P. Mohai (eds) *Race and the Incidence of Environmental Hazards: A Time for Discourse*, Boulder, CO: Westview Press.

Bullard, R. D. (1990) *Dumping in Dixie: Race, Class and Environmental Quality*, Boulder, CO: Westview Press.

Burgess, A. (2003) 'Review of P. Harremoes, D. Gee, M. MacGarvin, A. Stirling. J. Keys, B. Wynne and S. Guedes Vaz, *The Precautionary Principle in the 20th Century: Late Lessons from Early Warnings* (London: Earthscan, 2002)', *Health, Risk & Society*, 5(1): 105–7.

Burgess, J. and Harrison, C.M. (1993) 'The circulation of claims in the cultural politics of environmental change', in A. Hansen (ed.) *The Mass Media and Environmental Issues*, Leicester: Leicester University Press.

Burros, M. (2005) 'Stores say wild salmon, but tests say farm bred', *The New York Times*, 10 April.

Burstyn, V. (2005) *Water Inc.*, London: Verso.

Burton, A. (2003) 'Extinction of taxonomists hinders conservation', *Frontiers in Ecology*, 1(5): 231.

Bussey, J. (2005) 'Thailand tableau: rebuilt resorts, makeshift memorials', *The Wall Street Journal*, 23 June: A1.

Buttel, F. H. (1986) 'Sociology and the environment: the winding road toward human ecology', *International Social Science Journal*, 38(3): 337–56.

Buttel, F. H. (1987) 'New directions in environmental sociology', *Annual Review of Sociology*, 13: 465–88.

Buttel, F. H. (2000) 'Classical theory and contemporary environmental sociology: some reflections on the antecedents and prospects for reflexive modernization theories in the study of environment and society', in G. Spaargaren, A. P. J. Mol and F. H. Buttel (eds) *Environment and Global Modernity*, London: Sage.

Buttel F. H. (2002) 'Environmental sociology and the classical sociological tradition: some observations on current controversies' in R. E. Dunlap, F. H. Buttel, P. Dickens and A. Gijswijt (eds) *Sociological Theory and the Environment: Classical Foundations, Contemporary Insights*, Lanham, MD: Rowman & Littlefield.

Buttel, F. H. (2002) 'Has environmental sociology arrived?' *Organization & Environment*, 15(1): 42–54.

Buttel, F. H. (2003) 'Environmental sociology and the exploration of environmental reform', *Organization and Environment*, 16(3): 306–44.

Buttel. F. H. (2004) 'The treadmill of production: an appreciation, assessment, and agenda for research', *Organization & Environment*, 17(3): 323–36.

Buttel, F. H. and Taylor, P. (1992) 'Environmental sociology and global environmental change: a critical assessment', *Society and Natural Resources*, 5: 211–30.

Buttel, F. H. and Gijswijt, A. (2001) 'Environmental sociology: a retrospective on its first quarter century', in J. R. Blau (ed.) *Blackwell Companion to Sociology*, Oxford: Basil Blackwell.

Buttel, F. H. and Humphrey, C. R. (2002) 'Sociological theory and the natural environment', in R. E. Dunlap and W. Michelson (eds) *Handbook of Environmental Sociology*, Westport, CT: Greenwood Press.

Buttel, F.H., Dickens, P., Dunlap, R. E. and A. Gijswijt (2002) 'Sociological theory and the environment: an overview and introduction' in R. E. Dunlap, F. H. Buttel, P. Dickens and A. Gijswijt (eds) *Sociological Theory and the Environment: Classical Foundations, Contemporary Insights*, Lanham, MD: Rowman & Littlefield.

Cable, S. and Benson, M. (1993) 'Acting locally: environmental injustice and the emergence of grassroots environmental organizations', *Social Problems*, 40(4): 464–77.

Cable, S. and Cable, C. (1995) *Environmental Problems, Grassroots Solutions: The Politics of Grassroots Environmental Conflict*, New York: St. Martin's Press.

Calhoun, C. (2004) 'A world of emergencies: fear, intervention, and the limits of cosmopolitan order', *The Canadian Review of Sociology and Anthropology*, 41(4): 373–95.

Callon, M. (1986) 'Some elements of a sociology of translation: domestication of the scallops and the fishermen of St. Brieuc Bay', in J. Law (ed.) *Power, Action, and Belief: A New Sociology of Knowledge?*, London: Routledge & Kegan Paul.

Cantrill, J. G. (1992) 'Understanding environmental advocacy: interdisciplinary research and the role of cognition', *The Journal of Environmental Education*, 24: 35–42.

Capek, S. M. (1993) 'The "environmental justice" frame: a conceptual discussion and an application', *Social Problems*, 40: 5–24.

Capuzza, J. (1992) 'A critical analysis of image management within the environmental movement', *The Journal of Environmental Education*, 24: 9–14.

Carlton, J. (2005) 'Green groups, ranchers bury hatchet to curb oil drilling', *The Wall Street Journal*, 23 March: B1, B4.

Carolan, M. S. and Bell. M. M. (2004) 'No fence can stop it: debating dioxin drift from a small U.S. town to Arctic Canada', in N. E. Harrison and G. C. Bryner (eds) *Science and Politics in the International Environment*, Lanham, MD: Rowman & Littlefield.

Carson, R. (1962) *Silent Spring*, Boston, MA: Houghton Mifflin.

Carty, R. (2003) 'Whose hand on the tap? Water privatization in South Africa'. CBC Radio News, February. Online: www.cbc.ca/news/features/water/southafrica.html

Castells, M. (1996) *The Rise of the Network Society. Volume 1 of The Information Age: Economy, Society and Culture*, Malden, MA: Blackwell.

Castree, N. (2005) 'Review of Swyngedouw (2004)', *Urban Studies*, 42(8): 1471–2.

Catton, W. R. Jr. (1980) *Overshoot: The Ecological Basis of Revolutionary Change*, Urbana, IL: University of Illinois Press.

Catton, W. R. Jr. (1994) 'Foundations of human ecology', *Sociological Perspectives*, 37(1): 75–95.

Catton, W. R. Jr. (2002) 'Has the Durkheim legacy misled sociology?', in R. E. Dunlap, F. H. Buttel, P. Dickens and A. Gijswijt (eds) *Sociological Theory and the Environment: Classical Foundations, Contemporary Insights*, Lanham, MD: Rowman & Littlefield.

Catton, W. R. Jr. and Dunlap, R. E. (1978) 'Environmental sociology: a new paradigm', *The American Sociologist*, 13: 41–9.

Chavez, C. (1993) 'Farm workers at risk', in R. Hofrichter (ed.) *Toxic Struggles: The Theory and Practice of Environmental Justice,* Philadelphia, PA: New Society Publishers.

Christoforou, T. (2003) 'The precautionary principle and democratizing expertise: a European legal perspective', *Science and Public Policy*, 30(3): 205–11.

Churchill, W. and LaDuke, W. (1985) 'Radioactive colonization and the Native American', *Socialist Review*, 15: 95–120.

Ciccantell, P. S. (1999) 'It's all about power: the political economy and ecology of redefining the Brazilian Amazon', *The Sociological Quarterly*, 40(2): 293–315.

Cittadino, E. (1993) 'The failed promise of human ecology', in M. Shortland (ed.) *Science and Nature: Essays in the History of the Environmental Sciences*, Oxford: British Society for the History of Science.

Clapp, J. and Dauvergne, P. (2005) *Paths to a Green World: The Political Economy of the Global Environment*, Cambridge, MA: MIT Press.

Clarke, D. (1981) 'Second-hand news: production and reproduction at a major Ontario television station', in L. Salter (ed.) *Communication Studies in Canada*, Toronto: Butterworths.

Clarke, D. (1992) 'Constraints of television news production: the example of story geography', in M. Grenier (ed.) *Critical Studies in Canadian Mass Media*, Toronto: Butterworths.

Clarke, L. (1988) 'Explaining choices among technological risks', *Social Problems*, 35(1): 22–35.

Clarke, L. (2003) 'Conceptualizing responses to extreme events: the problem of panic and failing gracefully', *Research in Social Problems and Public Policy (Terrorism and Disaster: New Threats, New Ideas)*, Volume 11: 123–41.

Clarke, L. and Short, J. F. Jr. (1993) 'Social organization and risk: some current controversies', *Annual Review of Sociology*, 19: 375–99.

Clements, F. E. (1905) *Research Methods in Ecology*, Lincoln, NB: University Publishing Company, reprinted by Arno Press (1977).

Cline, W. R. (1992) *The Economics of Global Warming*, Washington, DC: Institute for International Economics.

Cohen, J. L. (1985) 'Strategy or identity: new theoretical paradigms and contemporary social movements', *Social Research*, 52: 663–716.

Collingridge, D. and Reeve, C. (1986) *Science Speaks to Power: The Role of Experts in Policy Making*, London: Frances Pinter.

Commoner, B. (1971) *The Closing Circle*, New York: Bantam Books.

Conklin, B. and Graham, L. (1995) 'The shifting middle ground: Amazonian Indians and eco-politics', *American Anthropologist*, 97(4): 695–710.

Corbett, J. B. (1993) 'Atmospheric ozone: a global or local issue? Coverage in Canadian and U.S. newspapers', *Canadian Journal of Communication*, 18: 81–7.

Corner, J. and Richardson, K. (1993) 'Environmental communication and the contingency of meaning: a research note', in A. Hansen (ed.) *The Mass Media and Environmental Issues*, Leicester: Leicester University Press.

Cotgrove, S. (1991) 'Sociology and the environment: Cotgrove replies to Newby', *Network* (British Sociological Association), 51, October.

Cottle, S. (1993) 'Mediating the environment: modalities of TV news', in A. Hansen (ed.) *The Mass Media and Environmental Issues*, Leicester: Leicester University Press.

Covello, V. T and Johnson, B. B. (1987) 'The social and cultural construction of risk: issues, methods and case studies', in B. B Johnson and V. T. Covello (eds) *The Social and Cultural Construction of Risk Selection and Perception*, Dordrecht, Holland: D. Reidel Publishing Company.

Cowling, E. B. (1982) 'Acid precipitation in historical perspective', *Environmental Science and Technology*, 16: 110A–123A.

Cracknell, J. (1993) 'Issues arenas, pressure groups and environmental agendas', in A. Hansen (ed.) *The Mass Media and Environmental Issues*, Leicester: Leicester University Press.

Crist, E. (2004) 'Against the social construction of nature and wilderness', *Environmental Ethics*, 26(1): 5–24.

Cronon, W. (1996) 'The trouble with wilderness; or getting back to the wrong nature', *Environmental History*, 1: 7–28.

Cronor, C. F. (1993) *Regulating Toxic Substances: A Philosophy of Science and Law*, New York: Oxford University Press.

Cylke, F. K. Jr. (1993) *The Environment*, New York: HarperCollins College Publishers.

Dake, K. (1992) 'Myths of nature: culture and social construction of risk', *Journal of Social Issues*, 48(4): 21–37.

Daley, P. and O'Neill, D. (1991) 'Sad is too mild a word: press coverage of the *Exxon Valdez* Oil Spill', *Journal of Communication,* 41: 42–57.

Davidson, D. J. and Frickel, S. (2004) 'Understanding environmental governance: a critical review', *Organization & Environment*, 17(4): 471–92.

Desfor, G. and Keil, R. (2004) *Nature and the City: Making Environmental Policy in Toronto and Los Angeles*, Tucson, AZ: The University of Arizona Press.

Dewar, E. (1995) *Cloak of Green*, Toronto: James Lorimer & Co.

Diamond, J. (2005) *Collapse: How Societies Choose to Fail or Succeed*, New York: Viking

Dickens, P. (1992) *Society and Nature: Towards a Greener Social Theory*, Hemel Hempstead: Harvester Wheatsheaf.

Dietz, T. R. and Rycroft, R. W. (1987) *The Risk Professionals*, New York: Russell Sage Foundation.

Dietz, T. R. *et al.* (2002) 'Risk, technology and society', in R. E. Dunlap and W. Michelson (eds) (2002) *Handbook of Environmental Sociology*, Westport, CT: Greenwood Press.

Dispensa, J. M. and Brulle, R. J. (2003) 'Media's social construction of environmental issues: focus on global warming – a comparative study', *International Journal of Environmental Issues*, 23(10): 74–105.

Dorfman, A. (1992) 'Sideshows galore', *Time*, 139(22), 1 June: 43.

Doughty, R. W. (1975) *Feather Fashions and Bird Preservation: A Study in Nature Protection,* Berkeley, CA: University of California Press.

Douglas, M. A. and Wildavsky, A. (1982) *Risk and Culture: An Essay on the Selection of Technological and Environmental Dangers*, Berkeley, CA: University of California Press.

Downey, G. L. (1986) 'Ideology and the Clamshell identity: organizational dilemmas in the Anti-Nuclear Power Movement', *Social Problems*, 33: 357–73.

Downs, A. (1972) 'Up and down with ecology – the "issue-attention" cycle', *The Public Interest*, 28: 38–50.

Doyle, J. (1985) *Altered Harvest: Agriculture, Genetics and the Fate of the World's Food Supply*, New York: Penguin.

Drexler, M. (2003) *Secret Agents: The Menace of Emerging Infections*, Toronto: Penguin.

Dryzek, J. S. (1997; 2nd edn 2005) *The Politics of the Earth: Environmental Discourses*, Oxford: Oxford University Press.

Dubet, F. (2004) 'Between a defence of society and a politics of the subject: the specificity of today's social movements', *Current Sociology*, 52(4): 693–716.

Dumoulin, D. (2003) 'Local knowledge in the hands of transnational networks: a Mexican viewpoint', *International Social Science Journal*, 178 (December): 593–605.

Duncan, O. D. (1961) 'From social system to ecosystem', *Sociological Inquiry*, 31: 140–9.

Dunk, T. (1994) 'Talking about trees: environment and society in forest workers' culture', *Canadian Review of Sociology and Anthropology* 31: 14–34.

Dunlap, R. E. (1993) 'From environmental problems to ecological problems', in C. Calhoun and G. Ritzer (eds), *Social Problems*, New York: McGraw Hill.

Dunlap, R. E. (2002a) 'Paradigms, theories and environmental sociology', in R. E. Dunlap, F. H. Buttel, P. Dickens and A. Gijswijt (eds) *Sociological Theory and the Environment: Classical Foundations, Contemporary Insights*, Lanham, MD: Rowman & Littlefield

Dunlap, R. E. (2002b) 'Environmental sociology: a personal perspective on its first quarter century', *Organization & Environment*, 15(1): 10–29.

Dunlap, R. E. and Catton, W. R. Jr. (1979) 'Environmental sociology', *Annual Review of Sociology*, 5: 243–73.

Dunlap, R. E. and Catton, W. R. Jr. (1992/3) 'Towards an ecological sociology: the development, current status and probable future of environmental sociology', *The Annals of the International Institute of Sociology*, 3 (New Series): 263–84.

Dunlap, R. E. and Michelson, W. (eds) (2002) *Handbook of Environmental Sociology*, Westport, CT: Greenwood Press.

Dunlap, R. E. and Rosa, E. A. (2000) 'Environmental sociology', in E. F. Borgatta and R. J. V. Montgomery (eds) *Encyclopedia of Sociology* (2nd edn, Vol. 2), New York: Macmillan.

Dunlap, R. E., Buttel, F. H., Dickens, P. and Gijswijt, A. (eds) (2002) *Sociological Theory and the Environment: Classical Foundations, Contemporary Insights*, Lanham, MD: Rowman & Littlefield.

Dunwoody, S. and Griffin, R. L. (1993) 'Journalistic strategies for reporting long-term environmental issues: a case study of three Superfund sites', in A. Hansen (ed.) *The Mass Media and Environmental Issues*, Leicester: Leicester University Press.

Durkheim, E. (2002) [1895] 'The rules of sociological method' in C. Calhoun, J. Gerteis, J. Moody, S. Pfaff, K. Schmidt and I. Virk (eds) *Classical Sociological Theory*, Malden, MA: Blackwell.

Dynes, R. R. (1970) *Organized Behavior in Disasters*, Lexington, MA: D. C. Heath and Co.

Eckersley, R. (2004) *The Green State: Rethinking Democracy and Sovereignty*, Cambridge, MA: MIT Press.

Edelman, M. (1964) *The Symbolic Use of Politics*, Urbana, IL: University of Illinois Press.

Edelman, M. (1977) *Political Language: Words That Succeed and Policies That Fail*, New York: Academic Press.

Eden, S., Tunstall, S. M. and Tapsell, S. M. (2000) 'Translating nature: river restoration as nature–culture', *Environment and Planning D: Society and Space*, 18(2): 257–73.

Eder, K. (1996) *The Social Construction of Nature: A Sociology of Ecological Enlightenment*, London: Sage.

Ehrlich, P. R. and Ehrlich, A. (1981) *Extinction: The Causes and Consequences of the Disappearance of Species*, New York: Random House.

Einsiedal, E. and Coughlan, E. (1993) 'The Canadian press and the environment: reconstructing a social reality', in A. Hansen (ed.) *The Mass Media and Environmental Issues*, Leicester: Leicester University Press.

Eisner, T. (1989–90) 'Prospecting for nature's chemical riches', *Issues in Science and Technology*, 6: 31–4.

Eisner, T. and Beiring, E. A. (1994) 'Biotic Exploration Fund: protecting biodiversity through chemical prospecting', *BioScience*, 44(2): 95–8.

Eldredge, N. (1992–3) 'Confronting the skeptics' (Review of E. O. Wilson, *The Diversity of Life* [Cambridge, MA: Harvard University Press, 1992]), *Issues in Science and Technology*, 9(2): 90–2.

Elliott, A. (2002) 'Beck's sociology of risk: a critical assessment', *Sociology* 36(2): 293–315.

Enloe, C. H. (1975) *The Politics of Pollution in a Comparative Perspective: Ecology and Power in Four Nations,* New York: David McKay.

Environmental Conservation (1993) 'Editor's note: the biodiversity treaty', *Environmental Conservation*, 20(3): 277.

Enzensberger, H. M. (1979) 'A critique of political ecology', in A. Cockburn and J. Ridgeway (eds) *Political Ecology*, New York: Quadrangle.

Escobar, A. (1996) 'Constructing nature: elements for a post-structural political ecology', in R. Peet and M. Watts (eds) *Liberation Ecology*, London: Routledge.

Etzioni, A. (1961) *A Comparative Analysis of Complex Organizations: On Power, Involvement and their Correlates*, Glencoe, IL: The Free Press.

Evans, D. (1992) *A History of Nature Conservation in Britain,* London: Routledge.

Evans, P. (2002) 'Introduction: looking for agents of urban livability in a globalized political economy', in P. Evans (ed.) *Livable Cities? Urban Struggles for Livelihood and Sustainability*, Berkeley, CA: University of California Press.

Eyerman, R. and Jamison, A. (1989) 'Environmental knowledge as an organizational weapon: the case of Greenpeace', *Social Science Information,* 28(1): 99–119.

Faber, D. and O'Connor, J. (1993) 'Capitalism and the crisis of environmentalism', in R. Hofrichter (ed.) *Toxic Struggles: The Theory and Practice of Environmental Justice*, Philadelphia, PA: New Society Publishers.

Faro, A. L. (2004) 'Actors, conflicts and the globalization movement', *Current Sociology*, 52(4): 633–47.

Ferree, M. M., Gamson, W. A., Gerhards, J. and Rucht, D. (2002) *Shaping Abortion Discourse: Democracy and the Public Sphere in Germany and the United States*, Cambridge: Cambridge University Press.

Finnegan, W. (2002) 'Leasing the rain: the race to control water turns violent', *The New Yorker*, 8 April: 43–53.

Firey, W. (1947) *Land Use in Central Boston*, Cambridge, MA: Harvard University Press.

Fisher, D. R. (2003) 'Global and domestic actors within the global climate change regime: toward a theory of the global environmental system', *International Journal of Sociology and Social Policy*, 23(10): 5–30.

Fishman, M. (1980) *Manufacturing the News*, Austin, TX: University of Texas Press.

Fletcher, F. J. and Stahlbrand, L. (1992) 'Mirror or participant? The news media and environmental policy', in R. Boardman (ed.) *Canadian Environmental Policy: Ecosystems, Politics and Process*, Toronto: Oxford University Press.

Forrest, T. R. (1973) 'Needs and group emergence: developing a welfare response' *American Behavioral Scientist,* 16(3): 413–25.

Foster, J. B. (1999) 'Marx's theory of metabolic rift: classical foundations for an environmental sociology', *American Journal of Sociology*, 105: 366–405.

Foucault, M. (1979) *Discipline and Punish: The Birth of the Prison*, trans. A. Sheridan, New York: Vintage.

Foucault, M. (1980) *Power/knowledge. Selected Interviews and Other Writings 1972–1977*, ed. C. Gordon, Brighton: Harvester.

Fox, S. R. (1981) *John Muir and His Legacy: The American Conservation Movement,* Boston, MA: Little, Brown & Co.

Franck, I. and Brownstone, D. (1992) *The Green Encyclopedia*, New York: Prentice Hall General Reference.

Frank, D. J. (1997) 'Science, nature and the globalization of the environment, 1870–1990', *Social Forces*, 76: 409–35.

Frank, D. J. (2002) 'The origins question: building global institutions to protect nature', in A. J. Hoffman and M. J. Ventresca (eds) *Organizations, Policy and the Natural Environment*, Stanford, CA: Stanford University Press.

Frank, D. J., Hironaka, A. and Schofer, E. (2000) 'The nation state and the natural environment over the twentieth century', *American Sociological Review*, 65, 96–116.

Freudenburg, W. R. (1997) 'Contamination, corrosion and the social order: an overview', *Current Sociology*, 45(3): 19–39.

Freudenburg, W. R. (2000) 'Social construction and social constrictions: toward analyzing the social construction of "the naturalized" as well as "the natural"', in G. Spaargaren, A. P. J. Mol and F. H. Buttel (eds) (2000) *Environment and Global Modernity*, London: Sage.

Freudenburg, W. R. (2001) 'Risk, responsibility and recreancy', in G. Böhm, J. Nerb, T. McDaniels and H. Speda (eds) *Research in Social Problems and Public Policy, Volume 9* ('Environmental Risks: Perception, Evaluation and Management'): 87–108.

Freudenburg, W. R. and Gramling, R. (1989) 'The emergence of environmental sociology: contributions of Riley E. Dunlap and William R. Catton, Jr.', *Sociological Inquiry*, 59(4): 439–52.

Freudenburg, W. R. and Jones, T. R. (1991) 'Attitudes and stress in the presence of technological risk', *Social Forces*, 69(4): 1143–68.

Freudenburg, W. R. and Pastor, S. (1992) 'Public responses to technological risks: toward a sociological perspective', *The Sociological Quarterly,* 33: 389–412.

Friedman, S. M. (1983) 'Environmental reporting: a problem child of the media', *Environment*, 25(10): 24–9.

Friedman, S. M. (1984) 'Environmental reporting before and after TMI', *Environment*, 26(10): 4–5: 34.

Gamson, W. A. and Modigliani, A. (1989) 'Media discourse and public opinion on nuclear power', *American Journal of Sociology*, 95: 1–37.

Gamson, W. A. and Wolfsfeld, G. (1993) 'Movements and media as interacting systems', *Annals* (AAPSS), 528: 114–25.

Gamson, W. A., Fireman, B. and Rytina, S. (1982) *Encounters With Unjust Authority*, Homewood, IL: The Dorsey Press.

Gamson, W. A., Croteau, D., Haynes, W. and Sasson, T. (1992) 'Media images and social construction of reality', *Annual Review of Sociology*, 18: 373–93.

Gans, H. J. (1979) *Deciding What's News: A Study of CBS Evening News, NBC Nightly News, Newsweek and Time,* New York: Pantheon Books.

Garrett, L. (1994) *The Coming Plague: Newly Emerging Diseases in a World Out of Balance*, New York: Penguin Books.

Gelcich, S., Edwards-Jones, G., Kaiser, M. J. and Watson, E. (2005) 'Using discourses for policy evaluation: the case of marine common property rights in Chile', *Society and Natural Resources*, 18: 377–91.

Gibbs, W. W. (2005) 'Obesity: an overblown epidemic?' *Scientific American,* 292(6): 70–77.

Giddens, A. (1981) *A Contemporary Critique of Historical Materialism*, Berkeley, CA: University of California Press.

Gitlin, T. (1980) *The Whole World Is Watching: Mass Media in the Making or Unmaking of the New Left,* Berkeley, CA: University of California Press.

Global Biodiversity Forum (1997) *Report of the Seventh Global Diversity Forum*, 6–8 June, Harare, Zimbabwe.

Goffman, E. (1974) *Frame Analysis: An Essay on the Organization of Experience*, Cambridge, MA: Harvard University Press.

Goldblatt, D. (1996) *Social Theory and the Environment*, Cambridge: Polity Press.

Goldman, M. and Schurman, R.A. (2000) 'Closing the "great divide": new social theory on society and nature,' *Annual Review of Sociology*, 26: 563–84.

Gooderham, M. (1994) 'Scientific group urges species inventory', *Globe & Mail* (Toronto), 23 February: A12.

Goodspeed, P. (2005) 'Violent clashes threaten tsunami relief', *National Post* (Toronto), 19 July: A14.

Gottlieb, R. (1993) *Forcing the Spring: The Transformation of the American Environmental Movement*, Washington, DC: Island Press.

Gould, K. A., Weinberg, A. S and Schnaiberg, A. (1993) 'Legitimating impatience: Pyrrhic victories of the modern environmental movement', *Qualitative Sociology,* 16(3): 207–46.

Greider, T. and Garkovitch, L. (1994) 'Landscapes: the social construction of nature and the environment', *Rural Sociology*, 59(1): 1–24.

Grumbine, R. E. (1992) *Ghost Bears: Exploring the Biodiversity Crisis*, Washington, DC: Island Press.

Gusfield, J. R. (1984) 'On the side: practical action and social constructivism in social problems theory', in J. I. Kitsuse (eds) *Studies in the Sociology of Social Problems,* Norwood, NJ: Ablex.

Guston, D. H. (2000) *Between Politics and Science: Assuring the Integrity and Productivity of Research*, New York: Cambridge University Press.

Haarmeyer, D. and Mody, A. (1997) 'Private capital in water and sanitation', *Finance and Development,* March, 34–7.

Haas, P. (1990) *Saving the Mediterranean: The Politics of International Environmental Cooperation,* New York: Columbia University Press.

Haas, P. (1992) 'Obtaining international protection through epistemic consensus', in I. H. Rowlands and M. Greene (eds) *Global Environmental Change and International Relations*, Basingstoke: Macmillan.

Hagen, J. B. (1992) *An Entangled Bank: The Origins of Ecosystem Ecology*, New Brunswick, NJ: Rutgers University Press.

Hajer, M. A. (1995) *The Politics of Environmental Discourse: Ecological Modernization and the Policy Process,* Oxford: Clarendon Press.

Hannigan, J. A. (1985) *Laboured Relations: Reporting Industrial Relations News in Canada*, Toronto: Centre for Industrial Relations, University of Toronto.

Hannigan, J. A. (1990) 'Emergence theory and the new social movements: a constructivist approach', paper presented at the World Congress of Sociology, Section on Collective Behavior and Social Movements, Madrid, 12 July.

Hannigan, J. A. (1995) *Environmental Sociology: A Social Constructionist Perspective*, London: Routledge.

Hannigan, J. A. (2002) 'Cultural analysis and environmental theory: an agenda', in R. E. Dunlap, F. H. Buttel, P. Dickens and A. Gijswijt (eds) *Sociological Theory and the Environment: Classical Foundations, Contemporary Insights*, Lanham, MD: Rowman & Littlefield.

Hannigan, J. A. (2005) 'The new urban political economy', in H. H. Hiller (ed.) *Urban Canada: Sociological Perspectives*, Toronto: Oxford University Press.

Hansen, A. (1991) 'The media and the social construction of the environment', *Media, Culture & Society*, 13(4): 443–58.

Hansen, A. (1993a) 'Introduction', in A. Hansen (ed.) *The Mass Media and Environmental Issues*, Leicester: Leicester University Press.

Hansen, A. (1993b) 'Greenpeace and press coverage of environmental issues', in A. Hansen (ed.) *The Mass Media and Environmental Issues*, Leicester: Leicester University Press.

Hardin, G. (1978) 'The tragedy of the commons', *Science*, 162: 1241–52.

Harrison, K. and Hoberg, G. (1991) 'Setting the environmental agendas in Canada and the United States: the case of dioxin and radon', *Canadian Journal of Political Science,* 24: 3–27.

Harrison, K. and Hoberg, G. (1994) *Risk, Science, and Politics: Regulating Toxic Substances in Canada and the United States,* Montreal: McGill-Queen's University Press.

Harrison, N. E. and Bryner, G. C. (2004) 'Towards theory', in N. E. Harrison and G. C. Bryner (eds) *Science and Politics in the International Environment*, Lanham, MD: Rowman & Littlefield.

Hart, D. M. and Victor, D. G. (1993) 'Scientific elites and the making of U.S. policy for climate change research, 1957–1974', *Social Studies of Science,* 23: 643–80.

Harvey, D. (1974) 'Population, resources and the ideology of science', *Economic Geography*, 50: 265–77.

Harvey, D. (1989) *The Condition of Postmodernity*, Oxford: Blackwell.

Harvey, F. (2004) 'Devastated environment will face long-term damage', *Financial Times* (London), 29 December: 2.

Hasegawa, K. (2002) 'Nobuko Iijima: "Mother of Environmental Sociology"', *Environment & Society* (Newsletter of Research Committee 24, International Sociological Association), No. 19 (January): 2.

Haughton, G. (2002) 'Market making: internationalization and global water markets', *Environment and Planning A*, 34(5): 791–807.

Hawkins, A. (1993) 'Contested ground: international environmentalism and global climate change', in R. D. Lipchutz and K. Conca (eds) *The State and Social Power in Global Environmental Politics,* New York: Columbia University Press.

Hays, S. (1959) *Conservation and the Gospel of Efficiency: The Progressive Conservation Movement*, Cambridge, MA: Harvard University Press.

Heimer, C. (1998) 'Social structure, psychology and the estimation of risk', *Annual Review of Sociology,* 14: 491–519.

Henninger, D. (2005) 'From spin city to fat city', *The Wall Street Journal*, 6 May: A14.

Herndl, C. G. and Brown, S. C. (1996) 'Introduction', in C. G. Herndl and S. C. Brown (eds) *Green Culture: Environmental Rhetoric in Contemporary America*, Madison, WI: University of Wisconsin Press.

Higgins, R. R. (1993) 'Race and environmental equity: an overview of the environmental justice issue in the policy process', *Polity,* 26(2): 281–300.

Higgins, V. and Natalier, K. (2004) 'Governing environmental harms in a risk society', in R. White (ed.) *Controversies in Environmental Sociology*, Cambridge: Cambridge University Press.

Higham, J. (1963) *Strangers in the Land: Patterns of American Nativism, 1860–1925*, New York: Atheneum.

Hilgartner, S. (1992) 'The social construction of risk objects: or how to pry open networks of risk', in J. F. Short Jr. and L. Clarke (eds) *Organizations, Uncertainties and Risk,* Boulder, CO: Westview Press.

Hilgartner, S. and Bosk, C. L. (1988) 'The rise and fall of social problems: a public arenas model', *American Journal of Sociology,* 94(1): 53–78.

Hindness, B. (2001) *Discourses of Power: From Hobbes to Foucault*, Oxford: Blackwell.

Hites, R. A., Foran, J. A., Carpenter, D. O., Hamilton, M. C., Knuth, B. A. and Schwager, S. J. (2004) 'Global assessment of organic contaminants in farmed salmon', *Science*, 303, (No.5655), 9 January: 226–9.

Hochberg, L. (1980) 'Environmental reporting in boomtown Houston', *Columbia Journalism Review*, 19: 71–4.

Hoffmann, J. (2004) 'Social and environmental influences on endangered species: a cross-national study', *Sociological Perspectives*, 47(1): 79–107.

Hofrichter, R. (1993) 'Introduction', in R. Hofrichter (ed.) *Toxic Struggles: The Theory and Practice of Environmental Justice*, Philadelphia, PA: New Society Publishers.

Hofstadter, R. (1959) *Social Darwinism in American Thought*, New York: George Braziller.

Horlick-Jones, T. (2004) 'Experts in risk? ... do they exist?', *Health, Risk & Society*, 6(2): 107–14.

Howard, L. E. (1953) *Sir Albert Howard in India,* London: Faber & Faber.

Howenstine, E. (1987) 'Environmental reporting: shift from 1970 to 1982', *Journalism Quarterly*, 64(4): 842–6.

Huber, J. (1982) *Die verlorene Unschuld der Ökologie: Neue Technologien und superindustrielle Entwicklung* (The Lost Innocence of Ecology: New Technologies and Superindustrial Development), Frankfurt am Main: Fischer.

Huber, J. (1985) *Die Regenbogengesellschaft: Ökologie und Sozialpolitik* (The Rainbow Society: Ecology and Social Policy), Frankfurt am Main: Fischer.

Humphrey, C. R., Lewis, T. L. and Buttel, F. H. (2003) 'Introduction: the development of environmental sociology', in C. R. Humphrey, T. L. Lewis and F. H. Buttel (eds) *Environment, Energy, and Society: Exemplary Works*, Belmont, CA: Wadsworth/Thomson Learning.

Hunt, S. A., Benford, R. D. and Snow, D. A. (1994) 'Identity fields: framing processes and the social construction of movement identities', in E. Larana, H. Johnston and J. R. Gusfield (eds) *New Social Movements: From Ideology to Identity,* Philadelphia, PA: Temple University Press.

Ibarra, P. R. and Kitsuse, J. I. (1993) 'Vernacular constituents of moral discourse: an interactionist proposal for the study of social problems', in J. A. Holstein and G. Miller (eds) *Reconsidering Social Constructionism: Debates in Social Problems Theory,* New York: Aldine de Gruyter.

Inkeles, A. and Smith, D. H. (1974) *Becoming Modern: Individual Change in Six Developing Countries*, Cambridge, MA: Harvard University Press.

Irwin, A. (2001) *Sociology and the Environment*, Cambridge: Polity Press.

Järvikowski, T. (1996) 'The relation of nature and society in Marx and Durkheim', *Acta Sociologica*, 39: 73–86.

Jasanoff, S. (1986) *Risk Management and Political Culture: A Comparative Study of Science in Policy Context,* New York: Russell Sage Foundation.

Jasanoff, S. (1990) *The Fifth Branch: Science Advisors as Policymakers*, Cambridge, MA: Harvard University Press.

Jenkins, J. C. (1983) 'Resource mobilization theory and the study of social movements', *Annual Review of Sociology*, 9: 527–53.

Jenness, V. (1993) *The Prostitutes' Rights Movement in Perspective,* Hawthorne, NY: Aldine de Gruyter.

Johnson, S. (2001) *Emergence: The Connected Lives of Ants, Brains, Cities, and Software*, New York: Scribner.

'Joining together for Justice' (2004) *Sierra*, 89(3) May–June: 57.

Jonassen, C. T. (1949) 'Cultural variables in the ecology of an ethnic group', *American Sociological Review*, 14: 32–41.

Kamenstein, D. S. (1988) 'Toxic talk', *Social Policy,* 19(2): 5–10.

Katz, E. and Lazarsfeld, P. (1955) *Personal Influence: The Part Played by People in the Flow of Mass Communication,* Glencoe, IL: Free Press.

Keane, J. and Mier, P. (1989) 'Editor's preface', in Melucci (1989).

Kebede, A. (2005) 'Grassroots environmental organizations in the United States: a Gramscian analysis', *Sociological Inquiry*, 75(1): 81–108.

Keck, M. E. (2002) 'Water, water, everywhere, nor any drop to drink: land use and water policy in São Paulo, Brazil', in P. Evans (ed.) *Livable Cities? Urban Struggles for Livelihood and Sustainability*, Berkeley, CA: University of California Press.

Kellert, S. R. (1986) 'Social and perceptual factors in the preservation of animal species', in B. G. Norton (ed.) *The Preservation of Species: The Value of Biological Diversity*, Princeton, NJ: Princeton University Press.

Kidner, D. W. (2000) 'Fabricating nature: a critique of the social construction of nature', *Environmental Ethics*, 22(4): 339– 57.

Killian, L. M. (1980) 'Theory of collective behavior: the mainstream revisited', in H. M. Blalock (ed.) *Sociological Theory and Research: A Critical Appraisal*, New York: The Free Press.

Killingsworth, M. J. and Palmer, J. S. (1992) *Ecospeak: Rhetoric and Environmental Politics in America*, Carbondale and Edwardsville, IL: Southern Illinois University Press.

Killingsworth, M. J. and Palmer, J. S. (1996) 'Millennial ecology: the apocalyptic narrative from *Silent Spring* to global warming', in C. G. Herndl and S. C. Brown (eds) *Green Culture: Environmental Rhetoric in Contemporary America*, Madison, WI: University of Wisconsin Press.

Kinchy, A. J. and Kleinman, D. L. (2003) 'Organizing credibility: Discursive and organizational orthodoxy on the borders of ecology and politics', *Social Studies of Science*, 33(6): 869–96.

Kingdon, J. W. (1984) *Agendas, Alternatives and Public Policies,* Boston, MA: Little, Brown and Co.

Kitsuse, J. I., Murase, A. E. and Yamamura, Y. (1984) 'The emergence and institutionalization of an educated problem in Japan', in J. W. Schneider and J. I. Kitsuse (eds) *Studies in the Sociology of Social Problems,* Norwood, NJ: Ablex.

Klausner, S. Z. (1971) *On Man in His Environment: Social Scientific Foundations of Research and Policy*, San Francisco, CA: Jossey-Bass.

Kowalok, M. (1993) 'Research lessons from acid rain, ozone depletion and global warming', *Environment,* 35(60): 13–20; 35–8.

Krimsky, S. (1979) 'Regulating recombinant DNA research', in D. Nelkin (ed.) *Controversy: Politics of Technical Decisions*, Beverly Hills, CA: Sage.

Krimsky, S. and Plough, A. (1988) *Environmental Hazards: Communicating Risks as a Social Process*, Dover, MA: Auburn House.

Krogman, N. T. and Darlington, J. (1996) 'Sociology and the environment: an analysis of journal coverage', *The American Sociologist*, 27: 39–55.

Kroll-Smith, S., Couch, S. R. and Marshall, B.K. (1997) 'Sociology, extreme environments and social change', *Current Sociology*, 45(3): 1–18.

Kunst, M. and Witlox, N. (1993) 'Communication and the environment', *Communication Research Trends,* 13: 1–31.

Kwa, C. (1993) 'Radiation ecology, systems and the management of the environment', in M. Shortland (ed.) *Science and Nature: Essays in the History of the Environmental Sciences*, Oxford: British Society for the History of Science.

Lacey, C. and Longman, D. (1993) 'The press and public access to the environment and development debate', *The Sociological Review,* 41(2): 207–43.

Larson, B. M. H., Nerlich, B. and Wallis, P. (2005) 'Metaphors and biorisks: the war on infectious diseases and invasive species', *Science Communication*, 26(3): 243–68.

Lash, S. and Wynne, B. (1992) 'Introduction' to U. Beck, *The Risk Society*, London: Sage.

Latour, B. (1993) *We Have Never Been Modern*, trans. C. Porter, Cambridge, MA: Harvard University Press.

Latour, B. (1999) *Pandora's Hope: Essays on the Reality of Science Studies*, Cambridge, MA: Harvard University Press.

Latour, B. (2000) 'When things strike back', *British Journal of Sociology*, 51(1): 107–23.

Lawson, S. (1994) 'Farm widow refights old pipeline foe', *Toronto Star*, 22 January: C–6.

Lear, L. K. (1993) 'Rachel Carson's *Silent Spring', Environmental History Review,* 17: 23–48.

Lee, C. (ed.) (1992) *Proceedings, First National People of Color Environmental Leadership Summit*, New York: United Church of Christ Commission for Racial Justice, December.

Lee, D. C. (1980) 'On the Marxian view of the relationship between man and nature', *Environmental Ethics*, 2: 3–16.

Lee, S. and Park, J. (2002) 'Environmental sociology in Korea', *Environment & Society* (Newsletter of Research Committee 24, International Sociological Association), No. 19 (January): 2–5.

Lerner, D. (1958) *The Passing of Traditional Society: Modernizing the Middle East*, Glencoe, IL: The Free Press.

Lewin, K. (1947) 'Group decision and social change', in T. M. Newcomb and E. L. Hartley (eds) *Readings in Social Psychology*, New York: Holt.

Lewis, T. L. and Humphrey, C. R. (2005) 'Sociology and the environment: an analysis of coverage in introductory sociology textbooks', *Teaching Sociology*, 33 (April): 154–69.

Liberatore, A. (1992) 'Facing global warming: the interactions between science and policy making in the European community', paper presented at the Symposium, 'Current Developments in Environmental Sociology', International Sociological Association Thematic Group on 'Environment and Society', Woudschoten, The Netherlands, June.

Lidskog, R. (1993) 'Review of U. Beck, *The Risk Society*' [London: Sage, 1992], *Acta Sociologica*, 36(4): 400–3.

Lidskog, R. (2001) 'The re-naturalization of society? Environmental challenges for sociology', *Current Sociology*, 49(1): 113–36.

Lindblom, C. E. and Cohen, D. H. (1979) *Usable Knowledge: Social Science and Social Problem Solving*, New Haven, CT: Yale University Press.

Lipschutz, R. D. (1996) *Global Civil Society and Global Environmental Governance: The Politics of Nature from Planet to Planet*, Albany: SUNY Press.

Lipschutz, R. D. (2004) *Global Environmental Politics: Power, Perspectives, and Practice*, Washington, DC: CQ Press.

Litfin, K. (1994) *Ozone Discourse*, New York: Columbia University Press.

Liu, P. L.-F., Lynett, P., Fernando, H., Jaffe, B. E., Fritz, H., Higman, B., Morton, R., Goff, J. and Synolakis, C. (2005) 'Observations by the International Tsunami Survey Team in Sri Lanka', *Science*, 308, 10 June: 159.

Lockie, S. (1997) 'Chemical risk and the self-calculating farmer: diffuse chemical use in Australian broadacre farming systems', *Current Sociology*, 45(3): 81–97.

Lockie, S. (2004) 'Social nature: the environmental challenge to mainstream social theory', in R. White (ed.) *Controversies in Environmental Sociology*, Cambridge: Cambridge University Press.

Louka, E. (2002) *Biodiversity and Human Rights: The International Rules for the Protection of Biodiversity*, Ardsley, NY: Transnational Publishers.

Lovejoy, T. E. (1986) 'Species leave the ark: one by one', in B. G. Norton (ed.) *The Preservation of Species: The Value of Biological Diversity*, Princeton, NJ: Princeton University Press.

Lovell, J. (2005) 'Kilimanjaro's global warming: Africa's tallest mountain stripped of snow a wake-up call to G8, environmentalist says', *Globe &Mail*, 15 March: A1, A16.

Lowe, P. D. and Goyder, J. (1983) *Environmental Groups in Politics*, London: Allen & Unwin.

Lowe, P. D. and Morrison, D. (1984) 'Bad news or good news: environmental politics and the mass media', *The Sociological Review,* 32(1): 75–90.

Luoma, J. (2004) 'The water thieves', *The Ecologist*, 34(2): 52–7.

Lupton, D. (2004) 'A grim health future: food risks in the Sydney press', *Health, Risk and Society,* 6(2): 187–200.

Lutts, R. H. (1990) *The Nature Fakers: Wildlife, Science and Sentiment*, Golden, CO: Fulcrum.

Lynch, B. D. (1993) 'The garden and the sea: U.S. Latino environmental discourses and mainstream environmentalism', *Social Problems*, 40(1): 108–24.

McComas, K. and Shanahan, J. (1999) 'Telling stories about global climate change', *Communication Research*, 26(1): 30–57.

Macdonald, D. (1991) *The Politics of Pollution: Why Canadians are Failing their Environment*, Toronto: McClelland & Stewart.

McDonald, K. (2004) 'Oneself as another: from social movement to experience movement', *Current Sociology*, 52(4): 575–93.

Macfarlane, R. (2005) '4x4s are killing my planet', *The Guardian*, 4 June: 6–7.

McIntosh, R. P. (1985) *The Background of Ecology: Concept and Theory*, Cambridge: Cambridge University Press.

McIntyre, S. (2005) 'Revisiting the stick', *National Post* (Toronto), 17 June: FP–19.

McManus, J. (ed.) (1989) 'Planet of the year', *Time*, 2 January.

McNeely, J. A. (1992) 'Review of R. Tobin, *The Expendable Future: U.S Politics and the Protection of Biological Diversity*' [Durham, NC: Duke University Press, 1991], *Environment,* 34(2): 25–7.

McNeely, J. A., Miller, K. R., Reid, W. V., Mittermeier, R. A. and Werner, T. B. (1990a) *Conserving the World's Biological Diversity*, Gland, Switzerland and Washington, DC: World Conservation Union, World Resources Institute, Conservation International, World Wildlife Fund, and World Bank.

McNeely, J. A., Miller, K. R., Reid, W. V., Mittermeier, R. A. and Werner, T. B. (1990b) 'Strategies for conserving biodiversity', *Environment,* 32(3): 16–20; 36–40.

Magill, F. N. (ed.) (1995) *Great Events From History II: Ecology and Environment Series, Vol. 1:1902–1944*, Pasadena, CA: Salem Press.

Mahon, T. (1985) *Charged Bodies: People, Power and Paradox in Silicon Valley*, New York: New American Library.

Manes, C. (1990) *Green Rage: Radical Environmentalism and the Unmaking of Civilization*, Boston, MA: Little, Brown & Co.

Mann, C. C. and Plummer, M. L. (1992) 'The butterfly problem', *The Atlantic Monthly*, 229(1): 47–70.

Mannion, A. (1993) 'Review of W. Reid *et al., Biodiversity Prospecting: Using Genetic Resources for Sustainable Development*' [Washington, DC: World Resources Institute, 1993], *Environmental Conservation,* 20(3): 286–7.

Martel, N. (2005) 'Eco-lessons taught in a surfer-girl patois', *The New York Times*, 28 March: B1–2.

Martell, L. (1994) *Ecology and Society: An Introduction*, Cambridge: Polity Press.

Marx, G. T. (1980) 'Conceptual problems in the field of collective behaviour', in H. M. Blalock (ed.), *Sociological Theory and Research: A Critical Appraisal*, New York: The Free Press.

Mayer, E. L. (1992) 'Environmental racism', *Audubon*, January/February: 30–2.

Mazur, A. and Lee, J. (1993) 'Sounding the global alarm: environmental issues in the U.S. national news', *Social Studies of Science*, 23: 671–720.

Meadows, D.H., Meadows D.L., Randers, J. and W.W. Behrens III (1972) *The Limits to Growth*, Washington, DC: Universe Books.

Melucci, A. (1989) *Nomads of the Present: Social Movements and Individual Needs in Contemporary Society*, Philadelphia, PA: Temple University Press.

Merchant, C. (1987) 'The theoretical structure of ecological revolutions', *Environmental Review,* 11(4): 265–74.

Mertig, A. G. and Dunlap, R. E. (2003) 'Environmentalism', in N. J. Smelser and P. B. Baltes (eds) *International Encyclopedia of the Social and Behavioral Sciences*, Kidlington: Elsevier Science.

Merton, R. H. and Nisbet, R. A. (1971) *Contemporary Social Problems,* 3rd edn, New York: Harcourt, Brace World.

Milbrath, L. W. (1989) *Envisioning a Sustainable Society: Learning Our Way Out*, Albany: SUNY Press.

Milgram, S. and Toch, H. (1969) Collective behavior and social movements', in G. Lindzey and E. Aronson (eds) *Handbook of Social Psychology*, 2nd edn, Vol. 4, Reading, MA: Addison Wesley.

Millar, C. I. and Ford, L. D. (1988) 'Managing for nature: from genes to ecosystems', *BioScience,* 38(7): 456–7.

Miller, C. (2001) 'Hybrid management: boundary organizations, science policy, and environmental governance in the climate regime', *Science, Technology & Human Values*, 26(4): 478–500.

Miller, C. A. and Edwards, P. N. (2001) 'Introduction: the globalization of climate science and climate politics', in C. A. Miller and P. N. Edwards (eds) *Changing the Atmosphere: Expert Knowledge and Environmental Governance*, Cambridge, MA: MIT Press.

Miller, V. D. (1993) 'Building on our past, planning our future: communities of color and the quest for environmental justice', in R. Hofrichter (ed.) *Toxic Struggles: The Theory and Practice of Environmental Justice*, Philadelphia, PA: New Society Publishers.

Mills, C. W. (1959) *The Sociological Imagination*, New York: Oxford University Press.

Milne, K. (1993) 'The perils of green pessimism', *New Scientist,* 138(1877), 12 June: 34–7.

Milton, K. (1991) 'Interpreting environmental policy: a social scientific approach', *Journal of Law and Society*, 18(1): 4–17.

Mitsuda, H. (2002) 'Conference report: Kyoto Environmental Sociology Conference (KESC2001), *Environment & Society* (Newsletter of Research Committee 24, International Sociological Association), No. 19 (January): 7–8.

Mitsuda, H. and Fisher, D. (2000) 'Environmental sociology in Japan', *Environment & Society* (Newsletter of Research Committee 24, International Sociological Association), No. 16 (July): 2–5.

Modavi, N. (1991) 'Environmentalism, state and economy in the United States', *Research in Social Movements, Conflicts and Change*, 13: 261–73.

Mol, A. P. J. (1997) 'Ecological modernization: industrial transformation and environmental reform', in M. Redclift and G. Woodgate (eds) *The International Handbook of Environmental Sociology*, London: Edward Elgar.

Mol, A. P. J. and Spaargaren, G. (2003) 'Towards a sociology of environmental flows: a new agenda for 21st century environmental sociology', paper presented at the International Conference on 'Governing Environmental Flows', Wageningen, The Netherlands, June 13–14.

Molotch, H. (1970) 'Oil in Santa Barbara and power in America', *Sociological Inquiry*, 40: 131–44.

Molotch, H. and Lester, M. (1975) 'Accidental news: the great oil spill as local occurrence and national event', *American Journal of Sociology,* 81(2): 235–60.

Monteiro, G. (2005) 'Review essay; discourses on abortion, discourses on politics: two studies in the politics of discourse,' *Current Sociology*, 53(1): 157–68.

Mooney, H. A. (2003) 'The ecology–policy interface', *Frontiers in Ecology and the Environment*, 1(1): 49.

Morrison, D. E., Hornback, H. E. and Warner, W. H. (1972) 'The environmental movement: some preliminary observations and predictions', in W. R. Burch Jr., N. H. Cheek Jr. and L. Taylor (eds.) *Social Behavior, Natural Resources and the Environment*, New York: Harper and Row.

Morse, S. (1993) *Emerging Viruses*, London: Oxford University Press.

Myerson, G. and Rydin, Y. (1996) *The Language of Environment: A New Rhetoric*, London: UCL Press.

Murdoch, J. (2001) 'Ecologising sociology: actor–network theory, co-construction and the problem of human exemptionalism', *Sociology*, 35(1): 111–33.

Murphy, P. and Maynard, M. (2000) 'Framing the genetic testing issue: discourse and cultural clashes among policy communities', *Science Communication*, 22(2): 133–53.

Murphy, R. (1994) *Rationality and Nature: A Sociological Inquiry into a Changing Relationship*, Boulder, CO: Westview Press.

Murphy, R. (2004) 'Disaster or sustainability: the dance of human agents with nature's actants', *The Canadian Review of Sociology and Anthropology*, 41(3): 249–66.

Nash, R. (1967) *Wilderness and the American Mind*, New Haven, CT: Yale University Press.

Nash, R. (1977) 'The value of wilderness', *Environmental Review*, 1: 14–25.

Nash, R. (1989) *The Rights of Nature: A History of Environmental Ethics*, Madison, WI: University of Wisconsin Press.

Nelkin, D. (1987) 'The culture of science journalism', *Society,* 24(6): 17–25.

Nelkin, D. (1989) 'Communicating technological risk: the social construction of risk perception', *Annual Review of Public Health*, 10: 95–113.

Norton, B. G. (1986) 'Introduction to Part 1', in B. G. Norton (ed.) *The Preservation of Species: The Value of Biological Diversity*, Princeton, NJ: Princeton University Press.

Noss, R. (1990) 'From endangered species to biodiversity', in K. Kohm (ed.) *Balancing on the Brink of Extinction*, Washington, DC: Island Press.

Novek, J. and Kampen, K. (1992) 'Sustainable or unsustainable development? An analysis of an environmental controversy', *Canadian Journal of Sociology*, 17: 249–73.

Oliver, J. E. (2005) *Obesity: The Making of an American Epidemic*, Oxford: Oxford University Press.

Oreskes, N. (2004) 'The scientific consensus on climate change', *Science*, 306 (No. 5702), 3 December: 1686.

Palmlund, I. (1992) 'Social drama and risk evaluation', in S. Krimsky and D. Golding (eds) *Social Theories of Risk*, Westport, CT: Praeger.

Papadakis, E. (1984) *The Green Movement in West Germany*, London: Croom Helm/St. Martin's Press.

Park, R. E. (1936) [1952] 'Human ecology', in R. E. Park (ed.) *Human Communities: The City and Human Ecology*, New York: The Free Press (originally published in 1936 in *American Journal of Sociology*, 42: 1–15).

Parlour, J. W. and Schatzow, S. (1978) 'The mass media and public concern for environmental problems in Canada: 1960–1972', *International Journal of Environmental Studies,* 13: 9–17.

Parsons, H. L. (1977) *Marx and Engels on Ecology*, Westport, CT: Greenwood Press.

Parsons, T. (1978) *Action Theory and the Human Condition*, New York: The Free Press.

Paterson, M. (2000) 'Swampy fever: media constructions and direct action politics', in B. Seel, M. Paterson and B. Doherty (eds) *Direct Action: British Environmentalism*, London: Routledge.

Pauw, J. (2005) 'Metered to death: how a water experiment caused riots and a cholera epidemic', Washington, DC: The Center for Public Integrity, July 23. Accessed at www.icij.org/water/report

Payne, B. A. and Cluett, C. (2002) 'Environmental sociology in nonacademic settings', in R. E. Dunlap and W. Michelson (eds) *Handbook of Environmental Sociology*, Westport, CT: Greenwood Press.

Pearce, F. (1991) *Green Warriors: The People and Politics Behind the Environmental Revolution*, London: The Bodley Head.

Pellow, D. N. (2000) 'Environmental inequality formation: toward a theory of environmental injustice', *American Behavioral Scientist*, 43(4): 581–601.

Peluso, N. L. (1992) *Rich Forests, Poor People: Resource Control and Resistance in Java*, Berkeley, CA: University of California Press.

Peluso, N. L. (2003) 'Territorializing local struggles for resource control', in P. Greenough and A. L. Tsing (eds) *Nature in the Global South: Environmental Projects in South and Southeast Asia*, Durham, NC: Duke University Press.

Perrow, C. (1984) *Normal Accidents: Living with High Risk Technologies*, New York: Basic Books.

Perry, M. (1994) 'The toilet bowl they call Sydney Harbor', *Toronto Star,* 24 July: B–7.

Picou, J. S. and Gill, D. A. (2000) 'The Exxon Valdez disaster as localized environmental catastrophe: dissimilarities to risk society theory', in M. J. Cohen (ed.) *Risk in the Modern Age: Social Theory, Science and Environmental Decision-Making*, Basingstoke: Macmillan.

Plein, L. C. (1991) 'Popularizing biotechnology: the influence of issue definition', *Science, Technology & Human Values*, 16(4): 474–90.

Polletta, F. and Jasper, J. M. (2001) 'Collective identity and social movements', *Annual Review of Sociology*, 27: 283–305.

Ponting, J. R. (1973) 'Rumor control centers: their emergence and operations', *American Behavioral Scientist,* 16(3): 391–401.

Raffaelli, D. (2004) 'How extinction patterns affect ecosystems', *Science*, 306 (No. 5699): 1141–2).

Rafter, N. (1992) 'Claims-making and socio-cultural context in the first U.S. eugenics campaign', *Social Problems*, 35: 17–34.

Redclift, M. (1984) *Development and the Environmental Crisis: Red or Green Alternatives?* New York: Methuen.

Redclift, M. (1986) 'Redefining the environmental "crisis" in the South', in J. Weston (ed.) *Red and Green: The New Politics of the Environment*, London: Pluto Press.

Redclift, M. and Woodgate, G. (1994) 'Sociology and the environment: discordant discourse?' in M. Redclift and T. Benton (eds) *Social Theory and the Global Environment,* London: Routledge.

Redclift, M. and Woodgate, G. (eds) (1997) *The International Handbook of Environmental Sociology*, Cheltenham: Edward Elgar.

Regalado, A. (2005) 'Global warring: in climate debate, the "hockey stick" leads to a face-off', *The Wall Street Journal,* 14 February: A1; A13.

Reid, W. V. (1993–4) 'The economic realities of biodiversity', *Issues in Science and Technology* 10(2): 48–55.

Reid, W. V., Laird, S. A., Meyer, C. A., Gamez, R., Sittenfield, A., Jantzen, D. H., Gollin, M. A. and Juma, C. (eds) (1993) *Biodiversity Prospecting: Using Genetic Resources for Sustainable Development*, Washington, DC: World Resources Institute.

Renn, O. (1992) 'Concepts of risk: a classification', in S. Krimsky and D. Golding (eds) *Social Theories of Risk*, Westport, CT: Praeger.

Richardson, M., Sherman, J. and Gismondi, M. (1993) *Winning Back the Words: Confronting Experts in an Environmental Public Hearing*, Toronto: Garamond Press.

Rittenhouse, C. A. (1991) 'The emergence of premenstrual syndrome as a social problem', *Social Problems*, 38(3): 412–25.

Ronderos, M. T. (2004) 'Going it alone', *The Ecologist*, 34(2): 58–9.

Roth, C. E. (1978) 'Off the merry-go-round, on the escalator' in W. B. Stapp (ed.) *From Ought to Action in Environmental Education*, Columbus, OH: ERIE/SMEAC.

Roué, M. (2003) 'US environmental NGOs and the Cree. An unnatural alliance for the preservation of nature?', *International Social Science Journal*, 178 (December): 619–27.

Rubin, C. T. (1994) *The Green Crusade: Rethinking the Roots of Environmentalism,* New York: The Free Press.

Rule, J. B. (1989) Rationality and non-rationality in militant collective action', *Sociological Theory*, 7, 145–60.

Ryan, C. (1991) *Prime Time Activism*, Boston, MA: South End Press.

Rycroft, R. W. (1991) 'Environmentalism and science: politics and the pursuit of knowledge', *Knowledge: Creation, Diffusion, Utilization*, 13(2): 150–69.

Salter, L. (with the assistance of E. Levy and W. Leiss) (1988) *Science and Scientists in the Making of Standards*, Dordrecht: Kluwer Academic.

Sandman, P. M, Sachman, D. B., Greenberg, M. and Gochfeld, M. (1987) *Environmental Risk and the Press: An Exploratory Assessment*, New Brunswick, NJ: Transaction Books.

Sartre, J. P. (1975) *The Transcendence of the Ego: An Existentialist Theory of Consciousness*, New York: Farrar, Straus and Giroux.

Schama, S. (1996) *Landscape and Memory*, Toronto: Vintage Canada.

Schlesinger, P. (1978) *Putting 'Reality' Together: BBC News*, London: Constable,

Schmitt, P. J. (1990) *Back to Nature: The Arcadian Myth in Urban America*, Baltimore, MA: Johns Hopkins University Press.

Schnaiberg, A. (1980) *The Environment: From Surplus to Scarcity*, New York: Oxford University Press.

Schnaiberg, A. (1993) 'Introduction: inequality once more, with (some) feeling', *Qualitative Sociology*, 16(3): 203–6.

Schnaiberg, A. (2002) 'Reflections on my 25 years before the mast of the environment and technology section', *Organization & Environment*, 15(1): 30–41.

Schnaiberg, A. (2003) 'American environmental studies', in N. J. Smelser and P. B. Baltes (eds) *International Encyclopedia of the Social and Behavioral Sciences*, Kidlington: Elsevier Science.

Schnaiberg, A. and Gould, K. A. (1994) *Environment and Society*, New York: St. Martin's Press.

Schnaiberg, A., Pellow, D. and Weinberg, A. (2002) 'The treadmill of production and the environmental state', in A. P. J. Mol and F. H. Buttel (eds) *The Environmental State under Pressure*, JAI.

Schoenfeld, A. C. (1980) 'Newspapers and the environment today', *Journalism Quarterly*, 57: 456–62.

Schoenfeld, A. C., Meier, R. F. and Griffin, R. J. (1979) 'Constructing a social problem: the press and the environment', *Social Problems*, 27(1): 38–61.

Schrepfer, S. R. (1983) *The Fight to Save the Redwoods: A History of Environmental Reform 1917–1978*, Madison, WI: The University of Wisconsin Press.

Schumacher. E. F. (1974) *Small is Beautiful*, London: Abacus.

Scotland, R. (1994) 'Marketing model helps rate brand performance', *The Financial Post*, (Toronto), 1 December: 23.

Scott, J. (2001) *Power*, Cambridge: Polity/Blackwell.

Sears, P. B. (1964) 'Ecology as a subversive subject', *Bioscience,* 14(7): 11–13.

Seippel, O. (2002) 'Modernity, politics, and the environment: a theoretical perspective', in R. E. Dunlap, F. H. Buttel, P. Dickens and A. Gijswijt (eds) *Sociological Theory and the Environment: Classical Foundations, Contemporary Insights*, Lanham, MD: Rowman & Littlefield.

Selznick, P. (1960) *The Organizational Weapon*, Glencoe, IL: Free Press.

Shackley, S. (2001) 'Epistemic lifestyles in climate change modeling', in C. A. Miller and P. N. Edwards (eds) *Changing the Atmosphere: Expert Knowledge and Environmental Governance*, Cambridge, MA: MIT Press.

Shapiro, S. (1993) 'Rejoining the battle against noise pollution', *Issues in Science and Technology*, 9(3): 73–9.

Sheail, J. (1976) *Nature in Trust: The History of Nature Conservation in Britain*, Glasgow: Blackie.

Sheldrake, R. (2005) 'Listen to the animals: Why did so many animals escape December's tsunami?', *The Ecologist*, 35(2):18–20.

Shelford, V. (ed.) (1926) *Naturalist's Guide to the Americas*, Baltimore, MD: Williams & Wilkins Co.

Shenon, P. (1994) 'A Vietnamese goat is imperiled by fame', *The New York Times,* 29 November: A–6.

Shepard, P. (1969) 'Ecology and Man: a viewpoint', in P. Shepard and D. McKinly (eds) *The Subversive Science: Essays Toward an Ecology of Man*, Boston, MA: Houghton Mifflin.

Sherif, M. (1936) *The Psychology of Social Norms*, New York: Harper.

Shibutani, T. (1966) *Improvised News: A Sociological Study of Rumor*, Indianapolis, IN: Bobbs-Merrill.

Shiva, V. (1990) 'Biodiversity, biotechnology and profit: the need for a People's Plan to protect biological diversity', *The Ecologist,* 20(2): 44–7.

Shiva, V. (1993) 'Farmers' rights, biodiversity and international treaties', *Economic and Political Weekly* (Bombay), 28(4), 3 April: 555–60.

Shiva, V. (2005) 'The lessons of the tsunami', *The Ecologist*, 35(2): 21–3.

Shiva, V. and Holla-Bhar, R. (1993) 'Intellectual piracy and the neem tree', *The Ecologist,* 23(6), November/December: 223–7.

Shove, E. (1994) 'Sustaining developments in environmental sociology', in M. Redclift and T. Benton (eds) *Social Theory and the Global Environment*, London: Routledge.

Simonis, U. (1989) 'Ecological modernization of industrial society: the strategic elements', *International Social Science Journal*, 121: 347–61.

Slater, C. (2002) *Entangled Edens: Visions of the Amazon*, Berkeley, CA: University of California Press.

Smith, C. (1992) *Media and Apocalypse: News Coverage of the Yellowstone Forest Fires, Exxon Valdez Oil Spill and Loma Prieta Earthquake*, Westport, CT: Greenwood Press.

Smith, M. (1999) 'To speak of trees: social constructivism, environmental values, and the future of deep ecology', *Environmental Ethics*, 21(4): 359–76.

Smithson, M. (1989) *Ignorance and Uncertainty: Emerging Paradigms*, New York: Springer-Verlag.

Snow, D. A., Rocheford, E. B. Jr., Warden, S. H. and Benford, R. D. (1986) 'Frame alignment processes, micromobilization and movement participation', *American Sociological Review*, 51: 464–81.

Solesbury, W. (1976) 'The environmental agenda: an illustration of how situations may become political issues and issues may demand responses from government: or how they may not', *Public Administration*, 54: 379–97.

Sorokin, P. A. (1964) [1928] *Contemporary Sociological Theories: Through the First Quarter of the Twentieth Century*, New York: Harper & Row.

Soulé, M. and Kohm, K. A. (1989) *Research Priorities for Conservation Biology*, Washington, DC: Island Press.

Soulé, M. and Lease, G. (1995) (eds) *Reinventing Nature? Responses to Postmodern Deconstruction*, Washington, DC: Island Press.

Spaargaren, G. (2000) 'Ecological modernization theory and the changing discourse on environment and modernity', in G. Spaargaren, A. P. J. Mol and F. H. Buttel (eds) (2000) *Environment and Global Modernity*, London: Sage.

Spaargaren, G. and Mol, A. P. J. (1992) 'Sociology, environment and modernity: ecological modernization as a theory of social change', *Society and Natural Resources*, 5: 323–44.

Spaargaren, G., Mol, A. P. J. and F. H. Buttel (eds) (2000) *Environment and Global Modernity*, London: Sage.

Spector, M. and Kitsuse, J. I. (1973) 'Social problems: a reformulation', *Social Problems*, 20: 145–59.

Spector, M. and Kitsuse, J. I. (1977) *Constructing Social Problems*, Menlo Park, CA: Cummings.

Spencer, J. W. and Triche, E. (1994) 'Media constructions of risk and safety: differential framings of hazard events', *Sociological Inquiry,* 64(2): 199–213.

Spray, S. L. and McGlothlin, K. L. (2003a) 'Introduction', in S. L. Spray and K. L. McGlothlin (eds) *Loss of Biodiversity*, Lanham, MD: Rowman & Littlefield.

Spray, S. L. and McGlothlin, K. L. (2003b) 'The global challenge: concluding thoughts on the loss of biodiversity', in S. L. Spray and K. L. McGlothlin (eds) *Loss of Biodiversity*, Lanham, MD: Rowman & Littlefield.

Staggenborg, S. (1993) 'Critical events and the mobilization of the pro-choice movement', *Research in Political Sociology*, 6: 319–45.

Stallings, R. (1990) 'Media discourse and the social construction of risk', *Social Problems,* 37: 80–95.

Stocking, H. and Holstein, L. W. (1993) 'Constructing and reconstructing scientific ignorance: ignorance claims in science and journalism', *Knowledge: Creation, Diffusion, Utilization,* 15: 186–210.

Stocking, H. and Leonard, J. P. (1990) 'The greening of the press', *Columbia Journalism Review,* 29 (November/December): 37–44.

Stokstad, E. (2004) 'Salmon survey stokes debate about farmed fish', *Science*, 303 (No. 5655), 9 January: 154–5.

Sunderlin, W. D. (2003) *Ideology, Social Theory and the Environment*, Lanham, MD: Rowman & Littlefield.

Susskind, L. E. (1994) *Environmental Diplomacy: Negotiating More Effective Global Agreements*, New York: Oxford University Press.

Sutton, P. W. (2004) *Nature, Environment and Society*, Basingstoke: Palgrave Macmillan.

Suzuki, D. (1994a) 'Amazon's "Tarzan" plans epic swim', *Toronto Star,* 4 July: B–1.

Suzuki, D. (1994b) 'Study gives biodiversity a big boost', *Toronto Star,* 5 March: B–6.

Swyngedouw, E. (1999) 'Modernity and hybridity: nature, *regeneracionismo*, and the production of the Spanish waterscape, 1890–1930', *Annals of the Association of American Geographers,* 89(3): 443–65.

Swyngedouw, E. (2004) *Social Power and the Urbanization of Water*, Oxford: Oxford University Press.

Szasz, A. (1994) *Ecopoplism: Toxic Waste and the Movement for Environmental Justice*, Minneapolis, MN: University of Minnesota Press.

Tangley, L. (1988) 'Research priorities for conservation', *BioScience,* 38(7): 444–8.

Tansley, A. G. (1939) 'British ecology during the past quarter century: the plant community and the ecosystem', *Journal of Ecology,* 27: 513–30.

Tarrow, S. (1988) 'National politics and collective action: recent theory and research', *Annual Review of Sociology*, 14: 421–40.

Taylor, D. E. (2000) 'The rise of the environmental justice paradigm', *American Behavioral Scientist*, 43 (4): 508–80.

Teuber, E. B. (1973) 'Emergence and change of human relations groups', *American Behavioral Scientist,* 16(3): 391–401.

Thompson, M. (1991) 'Plural rationalities: the rudiments of a practical science of the inchoate', in J. Hansen (ed.) *Environmental Concerns: An Interdisciplinary Exercise*, London: Elsevier Applied Science.

Tierney, K. J. (1977) 'Emergent norm theory as theory: an analysis and critique of Turner's formulation', paper presented at the Annual Meeting of the North Central Sociological Association.

Tierney, K. J. (2003) 'Disaster beliefs and institutional interests: recycling disaster myths in the aftermath of 9–11', *Research in Social Problems and Public Policy* (Terrorism and Disaster: New Threats, New Ideas), Volume 11: 33–51.

Tierney, K. J. (2004) 'The 9/11 Commission and disaster management: little depth, less context, not much guidance', *Contemporary Sociology*, 34 (2): 115–20.

Timasheff, N. S. and Theodorson, G. A. (1976) *Sociological Theory: Its Nature and Growth*, New York: Random House.

Timmer, V. and Juma, C. (2005) 'Taking root: biodiversity conservation and poverty reduction come together in the tropics: lessons learned from the Equator Initiative', *Environment*, 47(4): 24–44.

Tolba, M. K. and El-Kholy, O. A. (1992) *The World Environment 1972–1992: Two Decades of Challenge*, London: Chapman & Hall.

Tuchman, G. (1978) 'Introduction: the symbolic annihilation of women by the mass media', in G. Tuchman, A. K. Daniels and J. Benet (eds) *Hearth and Home,* New York: Oxford University Press.

Turner, R. H. (1981) 'Collective behavior and resource mobilization as approaches to social movements: issues and continuities', *Research in Social Movements, Conflict and Change,* 4: 1–24.

Turner, R. H. and Killian, L. M. (1957; 3rd edn 1987) *Collective Behavior*, Englewood Cliffs, NJ: Prentice Hall.

Udall, J. R. (1991) 'Launching the natural ark', *Sierra* 76(5), September/October: 80–9.

Ungar, S. (1992) 'The rise and (relative) decline of global warming as a social problem', *The Sociological Quarterly*, 33: 483–501.

Ungar, S. (1994) 'Apples and oranges: probing the attitude behaviour relationship for the environment', *Canadian Review of Sociology and Anthropology*, 31(3): 288–304.

Urry, J. (2003) *Global Complexity*, Cambridge: Polity.

Valiverronen, E. (1999) 'Review of D. Takacs, *The Idea of Biodiversity: Philosophies of Paradise*' [Baltimore, MD: Johns Hopkins University Press, 1996], *Science Technology & Human Values*, 24(3) 404–7.

van Koppen, C. S. A. (1998) 'Resource, arcadia lifeworld: nature concepts in environmental sociology', in A. Gijswijt, F. Buttel, P. Dickens, R. Dunlap, A. Mol and G. Spaargaren (eds) *Sociological Theory and the Environment: Proceedings of the Second Woudschoten Conference (Part Two: Cultural and Social Constructivism)*, SISWO, University of Amsterdam.

VanWynsberghe, R. (2001) 'Organizing a community response to environmental injustice: Walpole Island's Heritage Centre as a social movement organization', *Research in Social Problems and Public Policy*, 8 ('The Organizational Response to Social Problems') ed. S. W. Hartwell and R. K. Schutt): 221–43.

Vidal, J. (2005a) 'Flagship Africa scheme collapses', *The Guardian*, 25 May: 1.

Vidal, J. (2005b) 'Public backlash over private water deals', *The Guardian*, 25 May: 4.

Vogel, S. (1995) *The Concept of Nature in Critical Theory*, Albany, NY: SUNY Press.

von Weizsacker, C. (1993) 'Competing notions of biodiversity', in W. Sachs (ed.) *Global Ecology: A New Arena of Political Conflict*, London: Zed Books.

Walker, J. L. (1981) 'The diffusion of knowledge, policy communities and agenda setting: the relationship of knowledge and power', in J. E. Tropman, M. J. Dluhy and R. M. Lind (eds) *New Strategic Perspectives on Social Policy*, New York: Pergamon Press.

Weber, M. (1946) [1918] 'Science as a vocation', in H. H. Gerth and C. W. Mills (trans. and ed.), *From Max Weber: Essays in Sociology*, New York: Oxford University Press.

Weber, M. (1978) [1922] *Economy and Society: An Outline of Interpretive Sociology*, Guenther Roth and Claus Wittich (eds), Berkeley, CA: University of California Press.

Wegner, E. (1998) *Communities-of-Practice: Learning, Meaning, Identity*, Cambridge: Cambridge University Press.

Weiner, C. L. (1981) *The Politics of Alcoholism: Building an Arena around a Social Problem*, New Brunswick, NJ: Transaction Books.

Weitzer, R. (1991) 'Prostitutes' rights in the United States: the failure of a movement', *Sociological Quarterly*, 32(1): 23–41.

Weller, J. M. and Quarantelli, E. L. (1973) 'Neglected characteristics of collective behavior', *American Journal of Sociology*, 79: 665–85.

Wenger, D. (1978) 'Community response to disaster: functional and structural alterations', in E. L. Quarantelli (ed.) *Disasters: Theory and Research*, London: Sage.

West, P. C. (1984) 'Max Weber's human ecology of historical societies', in V. Murvar (ed.) *Theory of Liberty, Legitimacy and Power*, Boston, MA: Routledge and Kegan Paul.

Westell, D. (1994) 'Urban treatment of sewage gets bad marks: study shows 17 cities fail to treat all waste', *The Globe and Mail,* 15 June: A–4.

Wharton, C. R. Jr. (1966) 'Modernizing subsistence agriculture', in M. Wiener (ed.) *Modernization: The Dynamics of Growth*, New York: Basic Books.

Wilkins, L. and Patterson, P. (1990) 'Risky business: covering slow-onset hazards as rapidly developing news', *Political Communication and Persuasion,* 7(1): 11–23.

Wilkinson, I. (2001) 'Social theories of risk perception: at once indispensable and insufficient', *Current Sociology*, 49(1): 1–22.

Williams, J. (1998) 'Knowledge, consequences, and experience: the social construction of environmental problems', *Sociological Inquiry*, 68(4): 476–97.

Wilson, E. O. (1986) 'Editor's foreword', in E. O Wilson (ed.) *Biodiversity*, Washington, DC: National Academy Press.

Wilson, E. O. (1994) *Naturalist*, Washington, DC: Island Press.

Woolgar, S. and Pawluch, D. (1985) 'Ontological gerrymandering', *Social Problems*, 32(3): 214–37.

World Environmental Journalists (2005) 'Water markets in China increasing through government's five year plan, crisis impending', 21 March.

Worster, D. (1977) *Nature's Economy: A History of Ecological Ideas*, Cambridge: Cambridge University Press.

Wynne, B. (1992) 'Risk and social learning: reification to engagement', in S. Krimsky and D. Golding (eds) *Social Theories of Risk*, Westport, CT: Praeger.

Wynne, B. (2002) 'Risk and environment as legitimatory discourses of technology: Reflexivity inside out?' *Current Sociology*, 50(3): 459–77.

Wynne, B. and Mayer, S. (1993) 'How science fails the environment', *New Scientist,* 138(1876), 5 June: 33–5.

Yearley, S. (1992) *The Green Case: A Sociology of Environmental Issues, Arguments and Politics*, London: Routledge.

Yearley, S. (2002a) 'Review of B. Latour, *Pandora's Hope: Essays on the Reality of Science Studies*' [Cambridge, MA: Harvard University Press, 1999], *Science, Technology and Human Values*, 27(1): 165–7).

Yearley, S. (2002b) 'The social construction of environmental problems: a theoretical review and some not-very-Herculean labors', in R. E. Dunlap, F. H. Buttel, P. Dickens and A. Gijswijt (eds) *Sociological Theory and the Environment: Classical Foundations, Contemporary Insights*, Lanham, MD: Rowman & Littlefield.

York, R., Rosa, E. A. and Dietz, T. (2003) 'A rift in modernity? Assessing the anthropogenic sources of global climate change with the STIRPAT model', *International Journal of Sociology and Social Policy*, 23(10): 31–51.

Zamiska, N. (2005) '"Tsunami lung" strikes survivors; doctors fear widespread cases', *The Wall Street Journal*, 23 June: D2.

Zinn, J. O. (2005) 'The biographical approach: a better way to understand behaviour in health and illness', *Health, Risk & Society*, 7(1):1–9.

Name index

Subject index